T0295340

Phytochemicals in Vegetables and their Therapeutic Properties

NEW INDIA PUBLISHING AGENCY
New Delhi-110 034

Phytochemicals in Vegetables and their Therapeutic Properties

C.K. Narayana
Principal Scientist & Former HoD
Division of Postharvest Technology
ICAR-Indian Institute of Horticultural Research
Bengaluru. Karnataka

CRC Press
Taylor & Francis Group
Boca Raton London New York

CRC Press is an imprint of the
Taylor & Francis Group, an **informa** business

NEW INDIA PUBLISHING AGENCY
New Delhi-110 034

First published 2022
by CRC Press
2 Park Square, Milton Park, Abingdon, Oxon, OX14 4RN

and by CRC Press
6000 Broken Sound Parkway NW, Suite 300, Boca Raton, FL 33487-2742

© 2022 New India Publishing Agency

CRC Press is an imprint of Informa UK Limited

The right of C.K. Narayana to be identified as author of this work has been asserted by him in accordance with sections 77 and 78 of the Copyright, Designs and Patents Act 1988.

Print edition not for sale in South Asia (India, Sri Lanka, Nepal, Bangladesh, Pakistan or Bhutan).

British Library Cataloguing-in-Publication Data
A catalogue record for this book is available from the British Library

Library of Congress Cataloging-in-Publication Data
A catalog record has been requested

ISBN: 978-1-032-15678-1 (hbk)
ISBN: 978-1-003-24530-8 (ebk)

DOI: 10.1201/9781003245308

Dedication to
The Guiding Spirit
The Almighty God

Foreword

Vegetables are essential part of diet that provide fibres, minerals, vitamins besides antioxidants in the form of biomolecules like carotenoids, anthocyanins, phenolic compounds, alkaloids and others. Role of biomolecules in human health is being researched more extensively in recent times than ever before. Change in food habits with lesser intake of fruits and vegetables along with sedentary life style is contributing to increase in lifestyle diseases and disorders like diabetes, hypertension, hypercholesterolemia, cardiovascular diseases and different types of cancers. Phytochemicals in the vegetables are said to remove toxins, scavenge free radicals, repair damaged DNA and thereby prevent cancers. With advancement in analytical techniques it has become possible to probe into the phytochemicals present in fruits and vegetables, elucidate their structure and function and study their role in prevention or mitigation of many of the Non Communicable Diseases (NCDs).

India is home to one of the ancient systems of medicine (Ayurveda) and yoga, which are now transforming lives by suggesting corrections in food habits and lifestyle. Indian civilization being one of ancient were first to identified, study and use plants and herbs in management and curing several diseases of man and animals. Many of these plants and herbs were made part of food in order to keep body and mind healthy. Now we are witnessing a resurgence of belief in the complementarity of nutrition and pharmacology. Present nutritional research is coming out with scientific evidence and explanations about the role of plants and herbs in disease prevention and management, many of which are fruits and vegetables. This book on **'Phytochemicals in Vegetables and their Therapeutic Properties'**, provides compiled information on various nutritional aspects of commonly consumed vegetables in India along with research findings done in different parts of world, which substantiates claims in our ancient system of Indian Medicine (Ayurveda) besides traditional knowledge and folklore practices.

I commend the efforts made by the author in compiling the information published at various sources and presenting it in a comprehendible format for common man, students and researchers. I believe that the content of this book besides giving an insight into the benefits of vegetables as food and medicine, would incite the curious minds to probe it further and unravel the mysteries of nature.

Dr. G.G. Gangadharan
Director
M S Ramaiah Indic Centre for Ayurveda and
Integrative Medicine, Bengaluru, Karnataka, India

Preface

A comment published in The Lancet on 2nd February 2019, under the caption 'The 21st Century- great food transformation', says "for the first time in 200,000 years of human history, we are severely out of synchronization with the planet and nature. This crisis is accelerating, stretching Earth to its limits, and threatening human and other species' sustained existence".

According to ICMR, India: State-Level Disease Burden Study report, the estimated proportion of all deaths due to Non-communicable Diseases (NCDs) has increased from 37.09% in 1990 to 61.8% in 2016. India faces triple burden of health to deal with (infectious, NCDs and injuries). These figures indicate that not only the rich and affordable class suffer from lifestyle diseases, but even the poor and middle class too do. The comment in the Lancet further said, "The dominant diets that the world has been producing and eating for the past 50 years are no longer nutritionally optimal". The statement is loaded with message for the mankind.

The journey of the independent India, from a state of food deficiency to surplus was by and large, successful. With the shift of focus of governments to nutritional security, a kind of revolution could be achieved in production of nutritious foods, i.e., fruits, vegetables and nuts. Today, the food is just not for killing the hunger or provide calories, but fight diseases too. The current understanding of the disease development process has led the scientists to search for newer molecules from natural sources like fruits and vegetables. Several studies have proved that the food has been the cause of lifestyle diseases and it must be cured through food.

Several researchers have been studying, how to prevent or cure some serious diseases like cancer, CVD, AD, PD, diabetes and even infectious diseases through appropriate diet. This book (second in the series) on 'Phytochemicals in Vegetables & Their Therapeutic Properties' would provide the readers a bird's eye view of incredible value of vegetables in managing our health. As prevention is better than cure, let us know what we are eating and what it has in it.

For better health, vegetables should occupy a bigger space in meal, not just to fill the stomach and add nutrients, but also to keep at bay several day-to-day common ailments and prevent the disorders (lifestyle diseases) from occurring. During the compilation, efforts have been made, to the extent possible, to include traditional knowledge on health benefits of vegetables, as said in some scriptures and as believed in folklore.

This volume contains information on plant-based nutrients and phytochemicals in vegetable crops that we commonly consume. Also, the information generated by researchers using modern methods of biochemical analysis and, results mostly validated using cell line cultures or animal models, and to a limited extent on human volunteers have been presented. The vegetable crops have been grouped based on the family they belong to, because most often members of same family have similar biomolecules as active ingredients and have similar effects on health. References have been provided at the end of each chapter for further reading and better understanding of the subject. It is believed that this effort will help the researchers/scientists/students and common man and be able to appreciate the nature's marvel.

Any suggestions, inputs, information and assistance for improving it further would be highly appreciated.

April 2020
Bengaluru   **Author**

Acknowledgements

"Knowledge is in the end based on acknowledgement."
—Ludwig Wittgenstein

Behind every output of science, knowledge and technology, enough effort would have gone into igniting several minds with a spark of idea, inspiration, hand holding and guidance. The source of all these aspects for this particular book series has been Dr. H.P. Singh, Former Deputy Director General (Horticulture), Indian Council of Agricultural Research, New Delhi and Founder President, Confederation of Horticulture Associations of India, New Delhi. With all the reverence I acknowledge the kind guidance provided by him in bringing out two books namley (Phytochemicals in Fruits & Their Therapeutic Properties and Phytochemicals in Vegetables & Their Therapeutic Properties).

"It is easy to acknowledge, but almost impossible to realize
for long, that we are mirrors whose brightness, is wholly
derived from the sun that shines upon us."
—C.S. Lewis

I acknowledge with gratitude the readiness shown by Shri. Sumit Shobha Pal Jain and the team at New India Publishing Agency (NIPA), New Delhi for publishing these two books.

April, 2020   **C.K. Narayana**
Bengaluru

Contents

1. Introduction

Vegetables are important accompaniments for the main course of meal, which is by and large a carbohydrate-based food (wheat, rice or millets). It not only makes the meal more palatable, but also wholesome with its diverse color, smell and taste. More importantly it facilitates the easy movement of food through the bowel, better absorption by the body and till the elimination. Besides hunger satiation, it helps to maintain the health by providing the essential micronutrients and prevent lifestyle disorders.

As a nutritious food, it provides the vitamins, minerals and phyto-chemicals which are required in small quantities for physiological processes and metabolic activities. They are cheap sources of beta-carotene, vitamin-B complex, folic acid, vitamin C, vitamin E, *etc.* They are rich in minerals like iron, calcium, magnesium, phosphorus, *etc.*, and dietary fibres. It has now been proved that minerals are better absorbed by the body in presence of several other factors of plant matrix, than as mineral supplements. It also provides small amounts of carbohydrates and proteins. Vegetables are important as protective foods as their consumption prevents many diseases. They are important sources of antioxidants and phytochemicals. Each vegetable has been found to have certain specific phyto-chemical which besides being a free radical scavenger, is also inhibitor of certain conditions responsible for disease development.

Among the vegetables, tubers like cassava, potato and sweet potato being highly rich in carbohydrates, served as staples for several countries for very long time, in the history. For the same reason, there are a variety of food products based on these tubers. Others like carrot, radish, beetroot, and turnip in root vegetables, onion and garlic among bulbous crops; tomato, brinjal (egg plant), okra and capsicum among fruit vegetables serve as sources of phytochemicals (carotenoids, anthcyanins, lycopene, betalain, capsicum, bassiniloids, *etc.*). Beans, peas, cowpea, and other leguminous vegetables are important sources of proteins and leafy vegetables (greens) contribute to the much-needed minerals, vitamins and fibre (roughages) in the diet. Allicin and diallyl disulphide found in onion and garlic control blood cholesterol besides being anti-bacterial. The diphenylamine in onion is effective against diabetes. The 'Charantin' found in bitter gourd have hypoglycaemic activity. Diosgenin in yams are used for manufacture of contraceptive drugs. Cole crops like cabbage, cauliflower, Brussel sprouts and sprouting broccoli have anti-carcinogenic properties mainly due to hydrolysed glucosinolate derived isothiocynates and indoles. Similarly, health benefits and curative properties of many indigenous and exotic vegetables have been mentioned in several classical texts of India and China.

Transformation or shift in the dietary pattern of the world populations has brought mankind to the doorsteps of calamity, where non-infectious diseases and disorders have come to occupy the main spaces of medical research. Nutrition and epidemiological studies in the current period indicate the epidemic proportion of diseases and disorders originating from wrong food habits, besides dangerously changing lifestyles. The findings of research carried out by various laboratories across the world have been proving and validating these traditional claims using modern scientific tools and techniques. Every medical prescription of the day carries advise on food and exercise to supplement the corrective medicines.

Given this fact, it is essential that the vegetables have to be made available and accessible to the people of the country, in required quantity and quality to secure their health and reduce the expenditure due to disease burden. As the adage goes **'Prevention is better than cure'**, many of the common ailments can be prevented by consuming quality and wholesome food. It is recommended by the Indian Council of Medical Research (ICMR) that every individual should consume 300 grams of vegetables per day to be healthy. Against this backdrop, the current review of phytochemicals and therapeutic properties of vegetables shows what to eat for a better health and life.

A bird's eye view of our vegetable production in the country shows that, it has shown a phenomenal growth since 1991-92. In the decade between 1991-92 and 2001-02 though the area under cultivation of vegetables decreased marginally (2.34%), the production jumped by 31% showing its increased availability. Further in the subsequent decade, 2001-02 and 2010-11, there was a steep increase of 69% in the area under cultivation and the production shot up by 108%. In the year 2017-18, India produced 180.78 million tonnes of vegetables from 10.17 million ha. Impressive performance has been shown in area expansion, production and productivity by states like Andhra Pradesh, Gujarat, Tamil Nadu, Karnataka, Haryana, Chattisgarh and Rajasthan.

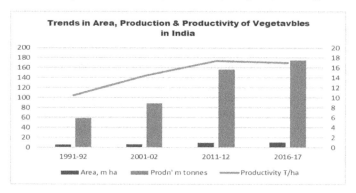

Creation of awareness about the importance of increased consumption of vegetables is one of the many approaches for nutritional security, a millinium developmental goal. In a humble effort in this direction, the wealth of information generated and available across the world on nutritional, therapeutic and curative properties of vegetables is made available in this volume for the benefit of readers, who could be public, students, researchers or policy makers. The category wise vegetables and their therapeutic uses validated through laboratory studies are summarized in the subsequent chapters.

2. Cucurbits

Cucurbits form an important and a big group of vegetable crops cultivated extensively in this country. This group consists of a wide range of vegetables, either used as salads (cucumber) or for cooking (all the gourds) or for pickling (cucumber) or as dessert fruits (muskmelon and watermelon) or candied or preserved (ash gourd). They are of tremendous economic importance as food plants.

2.1. Ash Gourd (*Benincasa hispida*)

Ash gourd (*Benincasa hispida* L.) belongs to the family Cucurbitaceae. The **winter melon**, also called **white gourd, ash gourd**, or "fuzzy melon", is a vine grown for its very large fruit, eaten as a vegetable when mature. It is the only member of the genus *Benincasa*. This is a monotypic genus and the only cultivated species is *Benincasa hispida* (Thunb) Cogn, commonly known as ash gourd or petha. It has a Sanskrit equivalent 'Kooshmanda' indicating its antiquity in India.

Originally cultivated in Southeast Asia, the ash gourd is widely grown in China, India and other parts of East Asia. The immature melon has thick white flesh that is sweet when eaten. By maturity, the fruit loses its hairs and develops a waxy coating, giving rise to the name wax gourd, and providing a long shelf life. The ash gourd is known by many other names in India regional wise like komora (Assamese), pethakaddu (Hindi), boodagumbala (Kannada), Kumbalanga (Malayalam), Kohla (Marathi), neer poosanikai (Tamil) and Boodida gummadikaya (Telugu).

Uses

Ash gourd have enormous medicinal properties as it makes a valuable place as a popular natural cure of ailments in traditional medicinal practice. Being low in calories it is an ideal vegetable for diabetic patients and good for weight control. It also acts as a good detoxifying agent.

As per Ayurveda, ash gourd is used to treat epilepsy, asthma, lung diseases, urine retention, internal haemorrhage, and cough. It is said that the fruit juice is also effective in cases of mercury poisoning and snakebite. It could also treat mouth cancer, protecting teeth and gums when a mouth gargle of the juice is done regularly. It is also effective in bleeding of gums. Ash gourd is easily digestible and keeps the body cool. It is a natural blood purifier.

Cultivars/Varieties: CO-1, CO-2, KAU Local and Indu are popular in Tamil Nadu and Kerala. Pusa Ujwal is very widely grown in plains of Northern India.

Nutritional Components of Ash gourd

Constituents	Quantity	Constituents	Quantity
Moisture	96.5 g/100 g	Calcium	30 mg/100 g
Carbohydrates	1.90 g/100 g	Phosphorus	20 mg/100 g
Protein	0.40 g/100 g	Iron	0.80 mg/100 g
Fat	0.10 g/100 g	Thiamine	0.06 mg/100 g
Minerals	0.30 g/100 g	Riboflavin	0.01 mg/100 g
Fibre	0.80 g/100 g	Niacin	0.40 mg/100 g
Energy	10 Kcal/100 g	Vitamin C	1 mg/100 g

(Gopalan *et al.*,1985)

Phytochemicals

Benincasa hispida has been shown to contain certain active principles like terpenes, flavonoid C – glycosides and sterols which have antioxidant effects (Shetty *et al.*, 2008). Through bioassay-guided separation, four known triterpenes and two known sterols were isolated from *Benincasa hispida*, as active components together with a flavonoid C-glycoside, an acylated glucose, and a benzyl glycoside. Among the active triterpenes and sterols, two triterpenes, alnusenol and multiflorenol, were found to potently inhibit the histamine release (Yoshizumi *et al.*, 1998). Raghu *et al.*, (2010) also reported presence of glycosides, alkaloids, flavonoids, tannins and polyphenols in *Benincasa hispida*.

Health Benefits of Ash gourd based on Traditional Knowledge/Folklore Practices

Benincasa hispida (Ash gourd) is a commonly used vegetable, which has found mention in 'Charaka Samhita' for its medicinal properties. It is the main ingredient in 'Kusumanda Lehyam" used in Ayurvedic system of medicine as a rejuvenating agent and in the treatment of nervous disorders. In Ayurveda, ash gourd is used to treat epilepsy, asthma, lung diseases, urine retention, internal haemorrhage, and cough. The fruit juice is also effective in cases of mercury poisoning and snakebites. It can even treat mouth cancer, protecting teeth and gums when a mouth gargle of the juice is done regularly. It is also effective in bleeding of gums.

- Being low in calories, it is ideal for diabetic patients, and those seeking weight-control. Due to its diuretic property it is believed to cool the body by promoting more urination. In the process it acts as a good detoxifying agent too.
- Ash gourd juice is recommended for those suffering from peptic ulcer. The juice derived from shredded ash gourd, when mixed with equal amount of water, should be taken every morning in an empty stomach, and thereafter no food should be consumed for three hours. Over a period of time, it is believed to cure peptic ulcer.
- Because ash gourd is a low acid vegetable, with pH almost near to neutral, it is believed to reduce stomach acidity and heart burn.
- Ash gourd has lots of fibres, due to which it provides good relief from constipation and tones up the general digestive system.

Therapeutic Properties of Ash Gourd

According to Kumar (2002), ash gourd is known to render protection against histamine induced bronchospasm. A multitude of studies have shown that the antioxidant status of an individual was compromised in most of these disorders. The fruit of *B hispida* is used as a diuretic and the seeds have been reported to possess anti-angiogenic effects in prostate cells.

Anti-ulcerogenic activity of different extracts of *B. hispida* (fresh juice, supernatant and residue fraction of centrifuged juice, alcoholic and petroleum ether extract) were studied in aspirin plus restraint, swimming stress, indomethacin plus histamine and serotonin-induced ulcers in rats and mice. The oral feeding of different doses of the extract significantly reduced the ulcer index produced by various ulcerogens (Grover *et al.*, 2001). Shetty *et al.*, (2008) studied the effect of ash gourd fruit extract on peptic ulcer and reported that terpene compounds were responsible for the antiulcerogenic effect and role of antioxidants present in *Benincasa hispida* has also been correlated. These compounds probably inhibit gastric mucosal injury by scavenging the indomethacin generated oxygen.

Manish *et al.*, (2008) opined that the mechanism of its gastroprotective activity may be attributed to reduction in vascular permeability, free radical generation, and lipid peroxidation along with strengthening of mucosal barrier. Besides, presence of phytoconstituents in this plant like flavone and sterols might be responsible for these actions.

Ash gourd has been found useful in prostrate related problems. Petroleum ether extract of *B. hispida* seed oil inhibited testosterone-induced hyperplasia of the prostate in rats (Nandecha *et al.*, 2010). Fruit has also been found useful in treating disorders of the GI tract, respiratory tract, urinary tract and diabetes mellitus (Aslokar *et al.*, 1992; Sivarajan and Balachandran, 1992).

References

Aslokar, L.V., Kakkar, K.K., and Chakre, O.J. 1992. Glossary, Indian medicinal plants with active principles. Part I, first edition, New Delhi, CSIR 1992; p. 119.

Gopalan, C.; B.V. Rama Sastri and S.C. Balasubramanian. 1985. Nutritive Value of Indian Foods. National Institue of Nutrition, ICMR, Hyderabad, India.

Grover, J. K., Adiga,G., Vats, V., and Rathi, S.S. 2001. Extracts of *Benincasa hispida* prevent development of experimental ulcers . Journal of Ethnopharmacology, 78(2-3): 159-164.

http://www.ayurvedictalk.com/medicinal-values-of-ash-gourd/1028/

Kumar, D., and Ramu, P. 2002. Effect of methanolic extract of *Benincasa hispida* against histamine and acetyl choline induced bronchospasm in guinea pigs. Indian J Pharmacol., (2002), 34: 365.

Manish, R.A., Jain, A., and Sunita, M. 2008. Gastroprotective effect of *Benincasa hispida* fruit extract. Indian Journal of Pharmacology, 40(6): 271-275.

Nandecha, C., Nahata,A., and Dixit, V.K. 2010. 2010. Effect of *Benincasa hispida* fruits on testosterone-induced prostatic hypertrophy in albino rats. Current Therapeutic Research, 71(5): 331-343.

Raghu, K.L., Ramesh, C.K., Srinivasa, T.R., and Jamuna, K.S. 2010. DPPH scavenging and reducing power properties in common vegetables. Research Journal of Pharmaceutical, Biological and Chemical Sciences, 1(4):399-406.

Shetty, B.V., Albina, A., Jorapur, A., Samanth, R., Yadav, S.K., Valliammai, N., Tharian, A.D., Sudha, K., and Ro, G.M. 2008. Effect of extract of *benincasa hispida* on oxidative stress in rats with indomethacin induced gastric ulcers. Indian J Physiol Pharmacol., (2008), 52(2) 178–182.

Sivarajan, V.V., Balachandran, I. (eds.) 1992. In: Ayurvedic drugs and their plant sources. 1st edition, New Delhi, CSIR, (1992): 119.

Yoshizumi, S., Murakami, T., Kadoya, M., Matsuda, H., Yamahara, J., Yoshikawa, M. 1998. Medicinal foodstuffs. XI. Histamine release inhibitors from wax gourd, the fruits of *Benincasa hispida* Cogn. Yakugaku Zasshi, 118(5):188-92. (Original in Japanese).

2.2. Bitter Gourd (*Mimordica charantia* L.)

Bitter gourd (*Mimordica charantia* L.) or balsam pear or *karela* is popular vegetable from cucurbitaceae family. It is a tropical and subtropical vine, widely grown in Asia, Africa and the Caribbean, for its edible fruit, which is among the most bitter of all fruits. Mimordica is a large genus comprising nearly 23 species in Africa alone. This genus is native to tropical regions of Asia with extensive distribution in China, Japan, South East Asia, Polynesia besides tropical Africa and South America. The cultivated species are *M. charantia*, the familiar bitter gourd is widely cultivated all over the country.

The bitter gourds of India have a narrower shape with pointed ends, and a surface covered with jagged, triangular "teeth" and ridges. It is green to white in color. Between these two extremes are any number of intermediate forms. Some bear miniature fruit of only 6–10 cm in length, which may be served individually as stuffed vegetable. This miniature fruit is popular in India and elsewhere in Southeast Asia.

Cultivars/Varieties: Pusa Do Mausmi, Coimbatore Long, Arka Harita, VK-1, Priya, Preethi, and Priyanka are some of the popular varieties which are long. Pride of Surat and Pride of Gujarat are small, roundish varieties which are whitish green in colour and weigh as less as 8-10 g.

Uses

The fruit is best known for medicinal value, the fruit though bitter is wholesome and considered as vegetable when physiologically mature. It is extensively used in ayurvedic formulations for a variety of ailments.

Nutritional Components of Bitter gourd

Constituents	Quantity (per 100g)	Constituents	Quantity (per 100g)
Moisture	92.4 g/100 g	Calcium	20 mg/100 g
Carbohydrates	4.20 g/100 g	Phosphorus	70 mg/100 g
Protein	1.60 g/100 g	Iron	1.80 mg/100 g
Fat	0.20 g/100 g	Thiamine	0.07 mg/100 g
Minerals	0.80 g/100 g	Riboflavin	0.09 mg/100 g
Fibre	0.80 g/100 g	Niacin	0.50 mg/100 g
Energy	25 Kcal/100 g	Vitamin C	88 mg/100 g

(Gopalan *et al.*, 1985)

Phytochemicals

The plant contains several biologically active compounds, chiefly momordicin I and II, and cucurbitacin B (Fatope *et al.*, 1990). The plant also contains several bioactive glycosides (including momordin, charantin, charantosides, goyaglycosides, momordicosides) and other terpenoid compounds including momordicin-28, momordicinin, momordicilin, momordenol, and momordol (Begum *et al.*, 1997; Okabe *et al.*, 1982; Kimura *et al.*, 2005; Chang *et al.*, 2008; Akihisa *et al.*, 2007). It also contains cytotoxic proteins (ribosome-inactivating) proteins such as momorcharin and momordin (Ortiao and Better, 1992). The total phenolic content of bitter gourd was reported as 42.36 mg GAE/g (Ibrahim *et al.*, 2010).

Main phenolic constituents in the water and ethanolic extracts of bitter gourd seeds were catechin, gallic acid, gentisic acid, chlorogenic acid, and epicatechin (Horax *et al.*, 2010). Gupta *et al.*, (2010) reported presence of Momordicatin, a purified extract obtained from bitter gourd. The leaves of bitter melon were found to contain nutrients like vitamin C, beta-carotene and lutein. The major flavonoids and phenolic acids were rutin, gentistic acid, and o-coumaric acid (Zhang *et al.*, 2009). Padmasree *et al.*, (2011) confirmed the presence of antioxygenic compounds in both bitter gourd pulp and seed. In particular, their ethanol/water extracts showed great potential as natural antioxidants to inhibit lipid peroxidation in foods.

Chen *et al.*, (2010) reported that two novel penta-norcucurbitane triterpenes, 22-hydroxy-23,24,25,26,27-pentanorcucurbit-5-en-3-one (1) and 3,7-dioxo-23,24,25,26,27-pentanorcucurbit-5-en-22-oic acid (2) together with a new trinorcucurbitane triterpene, 25,26,27-trinorcucurbit-5-ene-3,7,23-trione (3) were isolated from the methyl alcohol extract of the stems of *Momordica charantia*. The structures of the new compounds were elucidated by spectroscopic methods. Compounds 2 and 3 showed potent cytoprotective activity in tert-butyl hydroperoxide (t-BHP)-induced hepatotoxicity of HepG2 cells.

Hsu *et al.*, (2011) investigated the estrogenic activity and active cucurbitane-type triterpenoid compounds of bitter gourd (*Momordica charantia*, MC) using a transactivation assay for estrogen receptors (ER) α and β. The lyophilized fruits of *Momordica charantia* were exhaustively extracted with ethyl acetate (EA) and 95% ethanol (EtOH), sequentially. The non-saponifiable fraction (NS) of the EA extract as well as the acid hydrolyzed EtOH extract (AH) was fractionated and isolated by repeated column chromatography and further purified by preparative HPLC or RP-HPLC. One known compound, 5β,19-epoxycucurbita-6,24-diene-3β,23ξ-diol (6), was isolated from the NS, and five new compounds (1-5) were isolated from AH and identified as cucurbita-6,22I,24-trien-3β-ol-19,5β-olide (1), 5β,19-epoxycucurbita-6,22I,24-triene-3β,19-diol (2), 3β-hydroxycucurbita-5(10),6,22I,24-tetraen-19-al (3), 19-dimethoxycucurbita-5(10),6,22I,24-tetraen-3β-ol (4), and 19-nor-cucurbita-5(10),6,8,22I,24-pentaen-3β-ol (5).

Kimura *et al.*, (2005), isolated three new cucurbitane-type triterpenoids and established their structure as, (19R,23E)-5beta,19-epoxy-19-methoxycucurbita-6,23,25-trien-3beta-ol (1), (23E)-3beta-hydroxy-7beta-methoxycucurbita-5,23,25-trien-19-al (2), and (23E)-3beta-hydroxy-7beta,25-dimethoxycucurbita-5,23-dien-19-al (3) on the basis of spectroscopic methods.

Health Benefits of Bitter gourd based on Traditional Knowledge/ Folklore Practices

Bitter melon has been used in various Asian and African traditional medicine systems for a long time (Grover and Yadav, 2004; Nadine *et al.*, 2005; Paull *et al.*, 2010).

Digestive aid: Like most bitter-tasting foods, bitter melon is claimed to stimulate digestion, and thus help treat dyspepsia and constipation. However it is suspected of causing heartburn and ulcers, although these negative effects appear to be limited by its action as demulcent and mild inflammation modulator.

Antihelmintic: Bitter melon is used as a folk medicine in Togo to treat gastrointestinal diseases, and extracts have shown activity *in vitro* against the nematode worm *Caenorhabditis elegans* (Nadine *et al.*, 2005).

Antimalarial: Bitter melon is traditionally regarded in Asia as useful for preventing and treating malaria. Tea from its leaves is used for this purpose also in Panama and Colombia. In Guyana, bitter melons are boiled and stir-fried with garlic and onions. This popular side dish known as corilla is served to prevent malaria.

Antiviral: In Togo the plant is traditionally used against viral diseases such as chickenpox and measles. Tests with leaf extracts have shown *in-vitro* activity against the *Herpes simplex* type 1 virus, apparently due to unidentified compounds other than the momordicins. (Nadine *et al.*, 2005).

Anti-Diabetes: Ayurveda recommends Bitter Gourd as a general health tonic. It helps in maintaining blood sugar level and helps in purifying the blood. Its a natural laxative and helps in regulating the bowel movements. It is also possessing antibacterial properties and helps in skin ailments.

Other uses: Bitter melon has been used in traditional medicine for several other ailments, including dysentery, colic, fevers, burns, painful menstruation, scabies and other skin problems. It has also been used as abortifacient, for birth control, and to help childbirth.(Nadine *et al.*, 2005).

Therapeutic properties of Bitter gourd

Antidiabetic properties: Lolitkar and Rao (1962) extracted from the plant a substance, which they called charantin, which had hypoglycaemic effect on normal and diabetic rabbits. Bitter melon also contains a lectin that has insulin-like activity due to its non-protein-specific linking together to insulin receptors. This lectin lowers blood glucose concentrations by acting on peripheral tissues and, similar to insulin's effects in the brain, suppressing appetite. This lectin is likely a major contributor to the hypoglycemic effect that develops after eating bitter melon.laboratory studies have confirmed that species related to bitter melon have anti-malarial activity, though human studies have not yet been published (Waako *et al.*, 2005). Bitter gourd (*Momordica charantia*) contains substances with antidiabetic properties such as charantin, vicine, and polypeptide-p, as well as other unspecific bioactive components such as antioxidants. Metabolic and hypoglycemic effects of bitter gourd extracts have been demonstrated in cell culture, animal, and human studies, though the mechanism of action, whether it is via regulation of insulin release or altered glucose metabolism and its insulin-like effect, is still under debate. Some adverse effects have also been reported.

Nevertheless, bitter gourd has the potential to become a component of the diet or a dietary supplement for diabetic and pre-diabetic patients (Krawinkel and Keding, 2006). A study was carried out to examine the effect of edible portion of bitter gourd at 10% level in the diet in streptozotocin induced diabetic rats. To evaluate the glycaemic control of bitter gourd during diabetes, diet intake, gain in body weight, water intake, urine sugar, urine volume, glomerular filtration rate and fasting blood glucose profiles were monitored. Water consumption, urine volume and urine sugar were significantly higher in diabetic controls compared to normal rats and bitter gourd feeding alleviated this rise during diabetes by about 30%. Renal hypertrophy was higher in diabetic controls and bitter gourd supplementation, partially, but effectively prevented it (38%) during diabetes. Increased glomerular filtration rate in diabetes was significantly reduced (27%) by bitter gourd. An amelioration of about 30% in fasting blood glucose was observed with bitter gourd feeding in diabetic rats (Shetty et al., 2005).

Oxidative stress is currently suggested to play a major role in the development of diabetes mellitus. There is an increasing demand of natural anti-diabetic agents, as continuous administration of existing drugs and insulin are associated with many side effects and toxicity. Tripathi and Chandra (2009) investigated the effect of Momordica charantia (MC) and Trigonella foenum graecum (TFG) extracts (aqueous) on antioxidant status and lipid peroxidation in heart tissue of normal and alloxan induced diabetic rats. In a 30 days treatment, rats were divided into six groups (I-VI) of five animals in each, experiments were repeated thrice. Administration of MC (13.33 g pulp/kg body weight/day) and TFG (9 g seeds powder/kg body weight/day) extracts in diabetic rats has remarkably improved the elevated levels of fasting blood glucose. A significant decrease in lipid peroxidation and significant increase in the activities of key antioxidant enzymes such as superoxide dismutase (SOD), catalase (CAT), glutathione-s-transferase (GST) and reduced glutathione (GSH) contents in heart tissue of diabetic rats were observed (group V and VI) upon MC and TFG treatment. Lii et al., (2009) suggested that wild bitter gourd is beneficial for reducing LPS-induced inflammatory responses by modulating NF-kappaB activation which is also related with diabetes.

Anticancer: Two compounds extracted from bitter melon, α-eleostearic acid (from seeds) and 15,16-dihydroxy-α-eleostearic acid (from the fruit) have been found to induce apoptosis of leukemia cells in vitro (Miura et al., 2001). Kobori et al., (2008) found that diets containing 0.01% bitter melon oil (0.006% as α-eleostearic acid) prevented azoxymethane-induced colon carcinogenesis in rats.

Agrawal and Beohar (2010) demonstrated chemopreventive potential of Momordica fruit and leaf extracts on DMBA induced skin tumorigenesis, melanoma tumour and cytogenicity. Bitter melon (Momordica charantia Linnaeus) fruit extract was tested against 3,4 benzo(a)pyrene [B(a)P] induced forestomach papillomagenesis in Swiss albino mice. Extract of M. charantia in two concentrations, 2.5 and 5% of standard mice feed was used for the short-term and long-term studies. A significant decrease in tumour burden was observed in short and long-term treatment. Also, total tumour incidence reduced to 83.33% with 2.5% dose and 90.90% with 5% dose in short term treatment, while in long-term treatment tumor incidence decreased to 76.92% with 2.5% dose and 69.23% with 5% dose of M. charantia (Deep *et al.*, 2004).

The results of investigations carried out by (Pitchakarn *et al.*, 2011) suggest that bitter melon leaf extract and a purified component, Kuguacin J (KuJ), from its diethyl ether fraction could be promising candidate new antineoplastic and chemopreventive agents for androgen-dependent prostate cancer and carcinogenesis. They indicated for the first time an anti-metastatic effect of Bitter Melon Leaf Extract both *in vitro* and *in vivo*.

Laboratory tests suggest that compounds in bitter melon might be effective for treating HIV infection (Jiratchariyakul *et al.*, 2001). As most compounds isolated from bitter melon that impact HIV have either been proteins or lectins, neither of which are well-absorbed, it is unlikely that oral intake of bitter melon will slow HIV in infected people (Nerurkar *et al.*, 2006). The fruit is considered as tonic, stomachic, carminative and cooling and are used in rheumatic, gout and other diseases of liver and spleen (Kulkarni *et al.*, 2005).

References

Agrawal, R,C., and Beohar, T. 2010. Chemopreventive and anticarcinogenic effects of Momordica charantia extract. Asian Pac J Cancer Prev., (2010), 11(2):371-5.

Begum Sabira., Mansour Ahmed., Bina S. Siddiqui., Abdullah Khan., Zafar S. Saify., and Mohammed Arif. 1997. Triterpenes, a sterol, and a monocyclic alcohol from *Momordica charantia*. Phytochemistry, 44(7): 1313-1320.

Chang Chi-I., Chiy-Rong Chen., Yun-Wen Liao., Hsueh-Ling Cheng., Yo-Chia Chen and Chang-Hung Chou. 2008. Cucurbitane-type triterpenoids from the stems of *Momordica charantia*. Journal of Natural Products, 71(8): 1327–1330. doi:10.1021/np070532u.

Chen, C.R., Liao, Y.W., Wang, L., Kuo, Y.H., Liu, H.J., Shih, W.L., Cheng, H.L., Chang, C.I. 2010. Cucurbitane triterpenoids from *Momordica charantia* and their cytoprotective activity in tert-butyl hydroperoxide-induced hepatotoxicity of HepG2 cells. Chem Pharm Bull (Tokyo), (2010), 58(12):1639-42.

Deep, G., Dasgupta, T., Rao, A.R., Kale, R.K. 2004. Cancer preventive potential of *Momordica charantia* L. against benzo(a)pyrene induced fore-stomach tumourigenesis in murine model system. Indian J Exp Biol., (2004), 42(3):319-22.

Fatope, Majekodunmi., Yoshio, Takeda., Hiroyasu, Yamashita., Hikaru, Okabe., and Tatsuo, Yamauchi. 199., New cucurbitane triterpenoids from *Momordica charantia*. Journal of Natural Products, 53 (6):1491-1497.

Gadang, V., Gilbert, W., Hettiararchchy, N., Horax, R., Katwa, L., Devareddy, L. 2011. Dietary bitter melon seed increases peroxisome proliferator-activated receptor-γ gene expression in adipose tissue, down-regulates the nuclear factor-κB expression, and alleviates the symptoms associated with metabolic syndrome. J Med Food., (2011), 14(1-2):86-93.

Gopalan, C. B., V. Rama Sastri., and S.C. Balasubramanian. 1985. Nutritive Value of Indian Foods. National Institue of Nutrition, ICMR, Hyderabad, India.

Grover J. K., and Yadav, S. P. 2004. Pharmacological actions and potential uses of *Momordica charantia:* A Review. Journal of Ethnopharmacology, 93(1): 123–32. doi:10.1016/j. jep.2004.03.035

Gupta, S., Raychaudhuri, B., Banerjee, S., Das, B., Mukhopadhaya, S., Datta, S.C. 2010. Momordicatin purified from fruits of *Momordica charantia* is effective to act as a potent aphaninael agent. Parasitol Int., (2010), 59(2):192-7.

Horax, R., Hettiarachchy, N., Chen, P. 2010. Extraction, quantification, and antioxidant activities of phenolics from pericarp and seeds of bitter melons (*Momordica charantia*) harvested at three maturity stages (immature, mature, and ripe). J Agric Food Chem., (2010), 14;58(7):4428-33.

Hsu, C., Hsieh, C.L, Kuo, Y.H., Huang, C. J. 2011. Isolation and Identification of Cucurbitane-Type Triterpenoids with Partial Agonist/Antagonist Potential for Estrogen Receptors from *Momordica charantia*. J Agric Food Chem., (2011), 59(9):4553-61.

Ibrahim, T.A., El-Hefnawy, H.M., El-Hela, A.A. 2010. Antioxidant potential and phenolic acid content of certain cucurbitaceous plants cultivated in Egypt. Nat Prod Res., (2010), 24(16):1537-45.

Jiratchariyakul, W., Wiwat, C., Vongsakul, M., *et al*., 2001. HIV inhibitor from Thai bitter gourd. Planta Med., **67** (4): 350–3. doi:10.1055/s-2001-14323.

Kimura, Y., Akihisa, T., Yuasa, N., Ukiya, M., Suzuki, T., Toriyama, M., Motohashi, S., Tokuda, H. 2005. Cucurbitane-type triterpenoids from the fruit of *Momordica charantia*. J Nat Prod., (2005), 68(5):807-9.

Kimura, Yumiko., Toshihiro, Akihisa., Noriko, Yuasa., Motohiko, Ukiya., Takashi, Suzuki., Masaharu, Toriyama., Shigeyasu, Motohashi., and Harukuni, Tokuda. 2005. Cucurbitane-type triterpenoids from the fruit of *Momordica charantia*. Journal of Natural Products, 68(5):807-809. doi:10.1021/np040218p.

Kobori, Masuko., Mayumi, Ohnishi-Kameyama., Yukari, Akimoto., Chizuko, Yukizaki and Mitsuru, Yoshida. 2008. A-Eleostearic Acid and Its Dihydroxy Derivative Are MajorApoptosis-Inducing Components of Bitter Gourd. Journal of Agricultural and Food Chemistry, 56(22): 10515-10520.

Krawinkel, M.B., and Keding, G.B. 2006. Bitter gourd (*Momordica Charantia*): A dietary approach to hyperglycemia. Nutr Rev., (2006), 64(7) Part 1):331-337.

Kulkarni, A.S., H.B. Patil and Mundada, C.G. 2005. Studies on effect of pretreatment on quality of dehydrated bitter gourd (*Mimordica Charantia* L.) Adit Journal of Engineering, 2(1): 31-33.

Lii, C.K., Chen, H.W., Yun, W.T., and Liu, K.L. 2009. Suppressive effects of wild bitter gourd (*Momordica charantia* Linn. Var. aphanina ser.) fruit extracts on inflammatory responses in RAW264.7 macrophages. J Ethnopharmacol., (2009), 122(2):227-33.

Lolitkar, M. M., and Rao, M.R. 1962. Note on a Hypoglycaemic Principle Isolated from the fruits of *Momordica charantia*. Journal of the University of Bombay, 29: 223-224.

Miura, T., Itoh, C., Iwamoto, N., Kato, M., Kawai, M., Park, S.R., Suzuki, I. 2001. Hypoglycemic activity of the fruit of the *Momordica charantia* in type 2 diabetic mice. J Nutr Sci Vitaminol (Tokyo*)* **47** (5): 340–4.

Nadine, Beloin., Messanvi, Gbeassor., Koffi, Akpagana., Jim, Hudson., Komlan de Soussa., Kossi, Koumaglo and ThorArnason, J. 2005. Ethnomedicinal uses of *Momordica charantia* (Cucurbitaceae) in Togo and relation to its phytochemistry and biological activity. Journal of Ethnopharmacology, 96(1-2):49-55. doi:10.1016/j.jep.2004.08.009.

Nerurkar, P.V., Lee, Y.K., Linden, E.H., *et al.*, 2006. Lipid lowering effects of Momordica charantia (Bitter Melon) in HIV-1-protease inhibitor-treated human hepatoma cells, HepG2. Br. J. Pharmacol., **148** (8): 1156–64. doi:10.1038/sj.bjp.0706821.

Okabe, H., Y. Miyahara., and T. Yamauci. 1982. Studies on the constituents of *Momordica charantia L.* Chemical Pharmacology Bulletin, 30(12): 4334-4340.

Ortigao, Marcelo., and Marc Better 1992. Momordin II, a ribosome inactivating protein from *Momordica balsamina*, is homologous to other plant proteins. Nucleic Acids Research, 20(17): 4662.

Padmashree, A., Sharma, G.K., Semwal, A.D., Bawa, A.S. 2011. Studies on the antioxygenic activity of bitter gourd (*Momordica charantia*) and its fractions using various *in vitro* models. J Sci Food Agric., (2011), 91(4):776-82. Doi: 10.1002/jsfa.4251.

Paul, A., and Raychaudhuri, S.S. 2010. Medicinal uses and molecular identification of two *Momordica charantia* varieties – A Review. Electronic Journal of Biology, 6(2): 43-51.

Pitchakarn, P., Suzuki, S., Ogawa, K., Pompimon, W., Takahashi, S., Asamoto, M., Limtrakul, P., Shirai, T. 2011. Induction of G1 arrest and apoptosis in androgen-dependent human prostate cancer by Kuguacin J, a triterpenoid from *Momordica charantia* leaf. Cancer Lett., (2011), 306(2):142-50.

Shetty, A.K., Kumar, G.S., Sambaiah, K., Salimath, P.V. 2005. Effect of bitter gourd (*Momordica charantia*) on glycaemic status in streptozotocin induced diabetic rats. Plant Foods Hum Nutr., (2005), 60(3):109-12.

Sridhar, M.G., Vinayagamoorthi, R., Suyambunathan, A.V., Bobby, Z., Selvaraj, N. 2008. Bitter gourd (*Momordica charantia*) improves insulin sensitivity by increasing skeletal muscle insulin-stimulated IRS-1 tyrosine phosphorylation in high-fat-fed rats. British Journal of Nutrition, 99 (04): 806.

Toshihiro, Akihisa., Naoki, Higo., Harukuni, Tokuda., Motohiko, Ukiya., Hiroyuki, Akazawa., Yuichi, Tochigi., Yumiko, Kimura., Takashi, Suzuki., and Hoyoku, Nishino. 2007. Cucurbitane-type triterpenoids from the fruits of Momordica charantia and their cancer chemopreventive effects. Journal of Natural Products, 70:1233-1239.

Tripathi, U.N., Chandra, D. 2009. The plant extracts of *Momordica charantia* and *Trigonella foenum-graecum* have anti-oxidant and anti-hyperglycemic properties for cardiac tissue during diabetes mellitus. Oxid Med Cell Longev., (2009), 2(5):290-6.

Visarat, N., and Ungsurungsie, M. 1981. Extracts from Momordica charantia L. Pharmaceutical Biology, volume 19, issue 2–3, pages 75–80.

Waako, P.J., Gumede, B., Smith, P., Folb, P.I. 2005. The *in vitro* and *in vivo* antimalarial activity of *Cardiospermum halicacabum* L. and *Momordica foetida* Schumch. Et Thonn. J Ethnopharmacol., 99 (1): 137–43.

Zhang, M., Hettiarachchy, N.S., Horax, R., Chen, P., Over, K.F. 2009. Effect of maturity stages and drying methods on the retention of selected nutrients and phytochemicals in bitter melon (*Momordica charantia*) leaf. J Food Sci., (2009) 74(6):C441-8.

2.3. Bottel Gourd (*Lagenaria siceraria* Standl.)

Bottle gourd, or Lauki, belongs to the genus Lagenaria that has one cultivated monoecious species (*Lagenaria. Siceraria*-bottle gourd) and five wild perennial dioecious species. The dioecious species are confined to Africa and Madagascar. In India the fruits of bottle gourd show lot of variations in shape and size ranging from flattened or discoidal to bottle like forms.

Cultivars/Varieties: Pusa Summer Prolific Long, Pusa Summer Prolific Round, Pusa Meghdoot, Pusa Manjari are some varieties released by IARI, New Delhi

Uses

Lagenaria siceraria or *Lagenaria vulgaris*, also called the calabash, bottle gourd, opo squash or long melon is a vine grown for its fruit, which can either be harvested young and used as a vegetable, or harvested mature, dried, and used as a bottle, utensil, or pipe. They come in a variety of shapes, they can be huge and rounded, or small and bottle shaped, or slim and more than a meter long. For this reason, the calabash is widely known as the bottle gourd. The fresh fruit has a light green smooth skin and a white flesh. Rounder varieties are called Calabash gourds.

Hindu ascetics (sadhu) traditionally use a dried gourd vessel called the kamandalu. The juice of lauki is considered to have many medicinal properties and to be very good for health. The Baul singers of Bengal have their musical instruments made out of it. The practice is also common among Buddhist and Jain sages.

As, bottle gourd belongs to the family Cucurbitaceae it is also characterized by bitter principle called cucurbitacins. A systematic search for these substances in the family indicates that great majority of species contain bitter principles in some portion of the plant at some stage of development. Chemically cucurbitacins are tetracylclic triterpenes having extensive oxidation levels. They occur in nature, free as glycosides.

Nutritional Components of Ash gourd

Constituents	Quantity (per 100g)	Constituents	Quantity (per 100g)
Moisture	96.10 g/100 g	Calcium	20 mg/100 g
Carbohydrates	2.50 g/100 g	Phosphorus	10 mg/100 g
Protein	0.20 g/100 g	Iron	0.70 mg/100 g
Fat	0.10 g/100 g	Thiamine	0.03 mg/100 g
Minerals	0.50 g/100 g	Riboflavin	0.01 mg/100 g
Fibre	0.60 g/100 g	Niacin	0.20 mg/100 g
Energy	12 Kcal/100 g	Vitamin C	Nil

(Gopalan *et al.*,1985)

Phytochemicals

Ghosh *et al.*, (2009) reported that water-soluble polysaccharide, isolated from fruiting bodies of *Lagenaria siceraria*, is composed of methyl-alpha-d-galacturonate, 3-O-acetyl methyl-alpha-d-galacturonate, and beta-d-galactose in a ratio of nearly

1:1:1. This polysaccharide showed cytotoxic activity *in vitro* against human breast adenocarcinoma cell line (MCF-7).

Ghule *et al.*, (2006) conducted preliminary phytochemical screening of bottle gourd which revealed the presence of flavonoids, sterols, cucurbitacin, saponins, polyphenolics, proteins, and carbohydrates. Chen *et al.*, (2008) isolated four new D:C-friedooleanane-type triterpenes, 3 beta-O-I-feruloyl-D:C-friedooleana-7,9(11)-dien-29-ol (1), 3 beta-O-I-coumaroyl-D:C-friedooleana-7,9(11)-dien-29-ol (2), 3 beta-O-I-coumaroyl-D:C-friedooleana-7,9(11)-dien-29-oic acid (3), and methyl 2 beta,3 beta-dihydroxy-D:C-friedoolean-8-en-29-oate (6), together with five known triterpenes with the same skeleton, 3-epikarounidiol (4), 3-oxo-D:C-friedoolena-7,9(11)-dien-29-oic acid (5), bryonolol (7), bryononic acid (8), and 20-epibryonolic acid (9), from the methanol extract of the stems of *Lagenaria siceraria*.

Health Benefits of Bottle gourd based on Traditional Knowledge/Folklore practices

Bottle Gourd has many health benefits that are valued in traditional healing. Here are a few of them:

Low in Calories and Fat: The bottle gourd is low in fat (just 0.1 gm in 100 gm edible portion) and calories (15 calories in 100 gm) yet high in dietary fiber. It is comprised of 96% water and provides multiple vitamins (vitamin C and some B vitamins) and minerals (iron, sodium, potassium, and trace elements).

Cooling effect: As bottle gourd is mostly made of water, cooked bottle gourd is not only easy to digest but also contains cooling, calming and diuretic properties. A glass of bottle gourd juice taken in the morning can help prevent heat stroke during summers and cool the body.

Aids against constipation: Bottle gourd, being a good source of fiber content and easy to digest, tremendously helps in curing digestion related disorders like constipation, flatulence and piles.

Urinary disorders: A glass of fresh bottle gourd juice mixed with a teaspoon of lime juice serves as an alkaline mixture that cures the burning sensation in the urinary passage because of high acid content in urine.

Aids weight loss: Being low in calories and fat, and high in fiber and water makes bottle gourd an excellent food (both in cooked and juice form) to include in the diet when one is trying to maintain or lose weight.

Premature graying hair: According to Ayurveda, drinking a glass of bottle gourd juice in the morning daily can be very useful for treating graying hair.

Beneficial for skin: Bottle gourd in its natural form helps in internal cleansing of the skin. It helps to protect against various types of skin infections Including the condition of acne and other breakouts on the skin.

Insomnia: Mixture of bottle gourd juice and sesame oil acts as an effective medicine for insomnia. The cooked leaves of bottle gourd are also beneficial in the treatment of insomnia.

Revitalizes the body: A glass of bottle gourd juice is a valuable medicine for excessive thirst due to severe diarrhea, diabetes and excessive use of fatty or fried foods. It also prevents excessive loss of sodium, quenches thirst and helps in preventing fatigue.

Other ailments: Bottle gourd also has the capacity to cure disorders like toothaches, jaundice and inflammation of kidneys. *Lagenaria siceraria* (Mol.) is a commonly used vegetable in India is described as cardiotonic and as a general tonic in Ayurveda.

Therapeutic properties of Bottle gourd

Triterpenoids are structurally diverse organic compounds, characterized by a basic backbone modified in multiple ways, allowing the formation of more than 20,000 naturally occurring triterpenoid varieties. Several triterpenoids, including ursolic and oleanolic acid, betulinic acid, celastrol, pristimerin, lupeol, and avicins possess antitumor and anti-inflammatory properties. To improve antitumor activity, some synthetic triterpenoid derivatives have been synthesized, including cyano-3,12-dioxooleana-1,9 (11)-dien-28-oic (CDDO), its methyl ester (CDDO-Me), and imidazolide (CDDO-Im) derivatives. Of these, CDDO, CDDO-Me, and betulinic acid have shown promising antitumor activities. Triterpenoids are highly multifunctional and the antitumor activity of these compounds is measured by their ability to block nuclear factor-kappaB activation, induce apoptosis, inhibit signal transducer, and activate transcription and angiogenesis (Petronelli *et al.*, 2009).

Keeping in view the presence of free radical scavenging activity in *L. siceraria* and involvement of free radicals in the development of various disorders, Despande *et al.*, (2008) evaluated the ethanolic extract of *L. siceraria* fruit against the disorders where free radicals play a major role in pathogenesis. The extract was found effective as hepatoprotective, antioxidant, antihyperglycemic, immunomodulatory, antihyperlipidemic and cardiotonic agent. Their results showed that the radical scavenging capacity of *L. siceraria* fruit may be responsible for various biological activities studied.

A study undertaken by Ghule *et al.*, (2006) to explore the antihyperlipidemic effect of four different extracts viz. petroleum ether, chloroform, alcoholic and aqueous extracts from bottle gourd in Triton-induced hyperlipidemic rats and their hypolipidemic effects in normocholesteremic rats revealed that the chloroform and alcoholic extract exhibited more significant effects in lowering total cholesterol, triglycerides and low density lipoproteins along with increase in HDL.

Panda and Kar (2011) studied the effects of periplogenin-3- O-D-glucopyranosyl (1→6) (1→4)-D-cymaropyranoside, isolated from *Lagenaria Siceraria*, in L-thyroxine (L-T-)- on hyperthyroidism and its related cardiovascular abnormalities in Wistar albino rats. Their studies revealed protective role of periplogenin against thyrotoxicosis and associated cardiovascular problems. They were of the opinion that beneficial effects of periplogenin are mediated through its direct antithyroidal and/or (lipid peroxidation) LPO inhibiting properties.

References

Chen, C.R., Chen, H.W., Chang, C.I. 2008. D:C-friedooleanane-type triterpenoids from Lagenaria siceraria and their cytotoxic activity. Chem Pharm Bull (Tokyo). (2008), 56(3):385-8.

Deshpande, J.R., Choudhari, A.A., Mishra, M.R., Meghre, V.S., Wadodkar, S.G., Dorle, A.K. 2008. Beneficial effects of Lagenaria siceraria (Mol.) Standley fruit epicarp in animal models. Indian J Exp Biol., (2008), 46(4):234-42.

Ghosh, K., Chandra, K., Ojha, A.K., Sarkar, S., and Islam, S.S. 2009. Structural identification and cytotoxic activity of a polysaccharide from the fruits of Lagenaria siceraria (Lau). Carbohydr Res., (2009), 344(5):693-8.

Ghule, B.V., Ghante, M.H., Saoji, A.N., Yeole, P.G. 2006. Hypolipidemic and antihyperlipidemic effects of Lagenaria siceraria (Mol.) fruit extracts. Indian J Exp Biol., (2006), 44(11):905-9.

Gopalan, C., B.V. Rama Sastri and S.C. Balasubramanian. 1985. Nutritive Value of Indian Foods. National Institue of Nutrition, ICMR, Hyderabad, India.

Petronelli, A., Pannitteri, G., and Testa, U. 2009. Triterpenoids as new promising anticancer drugs. Anticancer Drugs., (2009), 20(10):880-92.

2.4. Pointed Gourd (*Trichosanthes dioica* Roxb.)

Parwal or Pointed gourd, as it is called belongs to the genus, Trichosanthes having more than 22 species, largely agreed to have been originated in India or Indo-Malayan region. *Trichosanthes dioica* is also known as the pointed gourd, parwal (from Hindi), or potol (from Assamese, Oriya or Bengali). Colloquially, in India, it is often called *green potato*. In pointed gourd there are four major types i) long dark green with white stripes, which may be 10-12 cm in length ii) 10-16 long with pale green colour and faint stripes iii) roundish dark green with stripes and as small as 5-8 cm and iv) small tapering at the ends, green and striped. These are mostly grown in Bihar, West Bengal, Eastern UP. It is widely cultivated in the eastern part of India, particularly in Orissa, Bengal, Assam, Bihar, Uttar Pradesh, and Madhya Pradesh.

Nutritional Components of Pointed gourd

Constituents	Quantity (per 100g)	Constituents	Quantity (per 100g)
Moisture	92.0 g/100 g	Calcium	30 mg/100 g
Carbohydrates	2.20 g/100 g	Phosphorus	40 mg/100 g
Protein	2.00 g/100 g	Iron	1.70 mg/100 g
Fat	1.10 g/100 g	Thiamine	0.05 mg/100 g
Minerals	0.50 g/100 g	Riboflavin	0.06 mg/100 g
Fibre	3.00 g/100 g	Niacin	0.50 mg/100 g
Energy	20 Kcal/100 g	Vitamin C	29 mg/100 g

(Gopalan *et al*., 1985)

It is a good source of carbohydrates, vitamin A, and vitamin C. It also contains major nutrients and trace elements (magnesium, potassium, copper, sulfur, and chlorine) which are needed in small quantities, for playing essential roles in human physiology (www. En.wikipedia.org/wiki/Trichosanthes_dioica).

Phytochemicals

Trichosanthes dioica revealed the presence of alkaloids, flavonoids, glycosides, tannins and steroids in different extracts of its fruits. These chemical constituents are responsible for the many therapeutic and medicinal properties of pointed gourd (Shivhare *et al*., 2010). Watal *et al*., (2010) reported that the higher concentration of Ca (2+), Mg (2+), and Fe (2+), are abundantly available in parval and are responsible for antioxidant potential of T. *dioica*. A new galactose-specific lectin has been purified from the extracts of *Trichosanthes dioica* seeds by affinity chromatography on cross-linked guar gum by Sultan *et al*., (2004). Lectins are carbohydrate-binding proteins that occur ubiquitously in nature and exhibit important biological properties such as bloodgroup specificity, preferential agglutination of tumour cells and mitogenicity (Lis & Sharon, 1991). The various chemical constituents present in TD are vitamin A, vitamin C, tannins, saponins, and trichosanthin (Raw materials, 2003; Chopra *et al*., 1956).

Leaves of the *Trichosanthes dioica* plant are considered to be rich source of bioactive compounds with many medicinal properties such as blood sugar lowering effect in experimental rat models (Chandra *et al*., 1988), mild diabetic human subject (Sharma *et al*., 1990), antifungal activity (Harit and Rathee, 1996) and antibacterial activity (Harit and Rathee, 1995). Phytochemical evaluations of aqueous and ethanolic extracts have showed the presence of saponins, tannins and a nonnitrogenous bitter glycoside trichosanthin (Khandelwal, 2005). Being very rich in protein and vitamin A, *Trichosanthes dioica* has certain medicinal properties.

Health Benefits of Pointed Gourd (Parval) Based on Traditional Knowledge/Folklore Practices

According to Ayurveda the plant is used for bronchitis, biliousness, cancer, jaundice, liver affections (Enlargement), cough, and blood diseases. It is also used as antipyretic diuretic, cardiotonic, laxative (Kirtikar and Basu, 1996). Free radicals have been implicated in the causation of several diseases such as liver cirrhosis, atherosclerosis, cancer, diabetes, *etc*., and compounds that can scavenge free radicals have great potential in ameliorating these disease processes. Antioxidants thus play an important role to protect the human body against damage by reactive oxygen species (Sabu and Kuttan, 2002).

Therapeutic Properties of Pointed Gourd

Trichosanthes dioica Roxb. (TD) commonly known as Kadu-padvala, is used in liver affections and jaundice (Kirtikar and Basu, 1996). TD extract was found to possess anti-inflammatory (Fulzule *et al*., 2001), cholesterol-lowering (Sharma and Pant, 1988; Sharmila *et al*., 2007), blood sugar, serum cholesterol, high density lipoprotein, phospholipids and triglyceride levels (Sharma and Pant, 1988; Chandrasekhar *et al*., 1988). The fruits are easily digestible and diuretic in nature. They are also known to have antiulcerous effects. The fruits and seeds have some prospects in the control of some cancer like conditions and haemagglutinating activities (Sharmila *et al*., 2007).

In glucose loaded rats, normal rats and hyperglycemic rats the aqueous extract of *Trichosanthes dioica* at both the doses (800 mg/kg/p.o and 1600 mg/kg/p.o) reduced blood glucose significantly when compared to control exhibiting its usefulness in treating hyperglycemia (Adiga *et al*., 2010). It also increases body weight of diabetic rats.

Tanwar *et al*., (2011) evaluated the hepatoprotective activity of ethanolic and aqueous extracts of *Trichosanthes dioica* Roxb. against paracetamol induced hepatic damage in rats. The substantially elevated serum levels of glutamate oxaloacetate transaminases (AST), glutamate pyruvate transaminases (ALT), alkaline phosphatase (ALP), total protein and total bilirubin were significantly restored by the test extracts. The results of the study indicated that *Trichosanthes dioica* Roxb. has hepatoprotective activity against paracetamol induced hepatic damage in rats. Aqueous extract was found to be more potent than ethanolic extract. *In vitro* antioxidant hydrogen peroxide (H_2O_2) and DPPH free radical scavenging activities were also screened which were positive for both ethanolic and aqueous extracts. The possible mechanism of this activity may be due to free radical-scavenging and antioxidant activities which may be due to the presence of saponins, tannins, vitamin C and carotene in the extracts. In rats with streptozotocin induced severe diabetes mellitus, aqueous extract of *Trichosanthes dioica* fruits at a dose of 1000mg/kg body weight daily once for 28 days reduced the levels of fasting blood glucose, postprandial glucose, asparate amino transferase, alanine amino transferase, alkaline phosphatase, creatinine, urine sugar and urine protein where as total protein and body weight was increased. No toxic effect was observed during LD50 (Rai *et al.*, 2006).

References

Adiga, Shalini., Bairy, K. L., Meharban, A., and Punita, I. S. R. 2010. Hypoglycemic effect of aqueous extract of *Trichosanthes dioica* in normal and diabetic rats. Int J Diabetes Dev Ctries., (2010), 30(1): 38–42.

Bose, T.K., and Som. M.G.1986. Vegetable crops in India. Naya Prokash, Calcutta.

Chandra, S., Mukherjee, B., Mukherjee, S.K. 1988. Blood sugar lowering effect of *Trichosanthes dioica* Roxb. In experimental rat models. Int. J. Crude Drug Res., 26(2): 102-106.

Chandrasekhar, B., Mukherjee, B., and Mukherjee, S.K. 1988. Blood sugar lowering effect of *Trichosanthes dioica* Roxb. In experimental rat models. Int. J. Crude Drug Res., 26: 102–106.

Chopra, R.N., Nayar, S.L., and Chopra, I.C. 1956. Glossary of Indian Medicinal plants, pp.256, CSIR, New Delhi.

Fulzule, S.V., Satturwar, P.M., Joshi, S.B. 2001. Studies on anti-inflammatory activity of a poly herbal formulation – Jatydi Ghrita. Indian Drugs, 39(1), 42-44.

Gopalan, C., B.V. Rama Sastri and S.C. Balasubramanian. 1985. Nutritive Value of Indian Foods. National Institue of Nutrition, ICMR, Hyderabad, India.

Harit, M., and Rathee, P.S. 1995. The antibacterial activity of the unsaponifiable fraction of the fixed oils of *Trichosanthes dioica* Seeds. Asian J. Chem., 7(4): 909-911.

Harit, M., and Rathee, P.S. 1996. The antifungal activity of the unsaponifiable fractious of the fixed oil of *Trichosanthes dioica* Roxb. Seeds. Asian J. Chem., 8(1):180-182.

Khandelwal, K.R. 2005. Practical Pharmacognosy, pp. 149–153 Nirali Prakashan, Pune.

Kirtikar, K.R., and Basu, B.D. 1996. Indian Medicinal Plants, pp. 1110-11, Jayyed Press, Allahabad.

Lis, H., and Sharon, N. (1991). Curr. Opin. Struct. Biol., 246, 227–234.

Rai, P. K., Jaiswal, D., Rai, D. K., Sharma, B., and Watal. G. 2006. Effect of water extract of *Trichosanthes dioica* fruits in streptozotocin induced diabetic rats. Indian Journal of Clinical Biochemistry, 23(4): 387-390.

Raw materials. 2003. The Wealth of India, pp. 289-291.

Sabu, M.C., Kuttan, R. 2002. Anti-diabetic activity of medicinal plants and its relationship with their antioxidant property. J. Ethnopharmacol., 81: 155-160.

Sharma, G., Pandey, D. N., Pant, M. C. 1990. The biochemical valuation of feeding *Trichosanthes dioica* Roxb. Seeds in normal & mild diabetic human subjects. In relation to lipid profile. Ind. J. Physiol. Pharmacol., 34(2):140-148.

Sharma, G. and Pant, M.C. 1988. Effect of raw deseeded fruit powder of *Trichosanthes dioica* (Roxb) on blood sugar, serum cholesterol, high density lipoprotein, phospholipids and triglyceride levels in the normal albino rabbits. Ind. J. Physiol. Pharmacol., 32(2): 161-163.

Sharma, G., Pant, M.C. 1988. Effects of feeding Trichosanthes dioica (parval) on blood glucose, serum triglyceride, phospholipids, cholesterol, and high-density lipoprotein-cholesterol levels in the normal albino rabbit. Current Science, 57: 1085– 1087.

Sharmila, B.G., Kumar, G., and Rajasekhar, P.M. 2007. Cholesterol-lowering activity of the aqueous fruit extract of *Trichosanthes dioica* Roxb. In normal and streptozotocin diabetic rats. J. of Clinical and Diagnostic Res., 1(4): 561-569.

Shivhare, Y., Singh, P., Shrivastava, S., Soni, P., Singha, A.K. 2010. Evaluation of Physicochemical & Phytochemical Parameters of *Trichosanthes dioica* Roxb. Herbal Tech Industry, December, 2010, Pages 17-19.

Sultan, N.A., Kenoth, R., and Swamy, M.J. 2004. Purification, physicochemical characterization, saccharide specificity, and chemical modification of a Gal/GalNAc specific lectin from the seeds of *Trichosanthes dioica*. Arch Biochem Biophys., (2004), 432(2):212-21.

Tanwar, M., Sharma, A., Swarnakar, K.P., Singhal, M., Yadav, K. 2011. Antioxidant and hepatoprotective activity of *Trichosanthes dioica* roxb. On paracetamol induced toxicity. International Journal of Pharmaceutical Studies and Research, II(I): 110-121.

Watal, G., Sharma, B., Rai, P.K., Jaiswal, D., Rai, D.K, Rai, N.K., and Rai, A.K. 2010. LIBS-based detection of antioxidant elements: a new strategy. Methods Mol Biol., (2010), 594:275-85.

2.5. Ivy Gourd or Kundru (Coccinia *indica*, L or *Coccinia grandis*, L)

Coccinia is a genus, in the family Cucurbitaceae that has nearly 30 speices confined to tropical Africa except *Coccinia grandis* (L), which extends throughout the tropics. *Coccinia grandis, L. or Coccinia indica* commonly called 'Ivy gourd' has a spread from Africa to Asia including India, Philippines, China, Indonesia, Malaysia, Thailand, Vietnam, eastern Papua, New Guinea and Northern Territories of Australia. It is extensively cultivated in India and is known by different names like Kundru (Punjabi), Tindori (Hindi), Dondakaya (Telugu & Kannada), Kovai or Kovakkai (Malayalam & Tamil) *etc.*, are a few common names in different states. This tropical dioecious species of India, called ivy gourd (kundru or tondli), like other vegetatively propagated cucurbits, is also very popular and extensively cultivated in the country. It has a Sanskrit equivalent name 'bimba' referring it to its existence in India since pre-Christian era.

In Tamil Nadu, Andhra Pradesh and Karnataka fruits are smaller, thinner and green in colour with longitudinal white stripes. In Western parts of the country the clones are medium sized (5 cm), long and thin with white stripes. In North and Eastern India, they are bigger and longer. Most of them are local clones propagated through cuttings.

Uses

It is primarily used for vegetable purpose.

Nutritionally some consider it one of the best among the cucurbits. Ivy plant has been used in traditional medicine as a household remedy for various diseases, including biliary disorders, anorexia, cough, diabetic wounds, and hepatic disorders.

Nutritional Components of Ivy gourd

Constituents	Quantity	Constituents	Quantity
Moisture (g/100g)	93.5	Calcium (mg/100g)	18
Carbohydrates (g/100g)	3.4	Phosphorus (mg/100g)	26
Protein (g/100g)	1.2	Iron (mg/100g)	0.5
Fat (g/100g)	0.1	Thiamine (mg/100g)	—
Minerals (g/100g)	0.5	Riboflavin (mg/100g)	0.01
Fibre (g/100g)	1.6	Niacin (mg/100g)	0.2
Energy (Kcal)	17	Vitamin C (mg/100g)	5.0

(Gopalan *et al.*, 1985)

Phytochemicals

For the last few decades, some extensive work has been done to establish the biological activities and pharmacological actions of Ivy Gourd and its extracts. Polyprenol (C60- polyprenol (1)) are the main yellow bioactive component of Ivy gourd and has been reported to have antidyslipidemic activity. Anti-inflammatory, antioxidant, antimutagenic, antidiabetic, antibacterial, antiprotozoal, antiulcer, hepatoprotective, expactorants, analgesi, antiinflamatory are the other pharmacological activities of Ivy Gourd reported (Yadav *et al.*, 2010).

The nutritional and phytochemical screening of coccinia using 50% methanolic extract obtained from whole parts of *Coccinia indica* revealed that it contains carbohydrates, glycosides, fixed oils and fats, proteins and amino acids as major nutritients. Saponins, tannins, phytosterol, alkaloids, phenolic compounds, flavonoids, gum and mucilage are the major phytochemicals (Shaheen *et al.*, 2009). Methanolic extract obtained from fruits of Ivy Gourd (*Coccinia indica*) were found to contain steroids, saponins, ellagic acid, lignins and triterpenoids, in addition to alkaloids, tannins, flavonoids, glycosides, phenols (Chandira, *et al.*, 2010). The aqueous extract of fresh leaves of Ivy Gourd (*Coccinia indica*) exhibited anthraquinons in addition to alkaloids, carbohydrate, proteins and amino acids, tannins, saponins, flavonoids, phytosterol, triterpenes (Niazi *et al.*, 2009). Presence of cephalandrol, tritriacontane, lupeol, b-sitosterol, cephalandrine A, cephalandrine B, stigma-7-en-3-one, taraxerone and taraxerol, have been reported by Rastogi *et al.*, (1998) and Ray and Kundu (1987).

Health Benefits of Ivy gourd based on Traditional Knowledge/Folklore practices

All parts of plant including leaf, fruit, root and stem were used in treating various kinds of diseases and disorders, which are listed below.

- Leaves were used in skin diseases (ring worm, psoriasis itch, sores, pityriasis), skin eruptions of smallpox, small lesions of scabies, chronic sinuses, gastro-intestinal disturbance and diseases.
- It alleviates body heat by inducing perspiration in fever. Tincture made out of leaves is used internally in gonorrhea.
- It also cures diabetes and intermittent glycosuria.
- It is sometimes used as expectorant, treating urinary tract infection and respiratory tract related trouble. The stem of coccinia has been found useful as antispasmodic agent, expectorant and in asthma and bronchitis.
- Like leaves & stem is also useful in controlling diabetes and intermittent glycosuria.
- The coccinia fruit cures sores on tongue and eczema.
- Roots remove pain in joints, cures apthous ulcers and wheezing.

Therapeutic properties of Ivy gourd (Kundru)

The pharmacological studies done using rat models indicated that the anti-diabetic effect of ivy gourd leaf extract could partly be due to the repression of the key gluconeogenic enzyme glucose-6-phosphatase (Hossain *et al.*, 1992). Kumar *et al.*, (1993) observed hypoglycemic effect in oral administration of pectin extracted from *Coccia indica* and opined that the reduced blood glucose and liver glycongen could be due to significant reduction in phosphorylase activity, in presence of some of the phytochemicals in *Coccia.*

A study was conducted to evaluate the antihyperglycaemic, antihyperlipidemic and antioxidative effects of aqueous methanolic extracts of the leaves of *Coccinia indica* separately as well as in composite manner with banana leaf extract, in streptozotocin-(STZ) induced diabetic male albino rats by Mallick *et al.*, (2007). The results revealed that co-administration of the composite extract of the above plant parts reversed the levels of fasting blood glucose, serum insulin and glycosylated haemoglobin (Ghb) towards the control level. Lipid peroxidation indicators such as serum levels of total cholesterol (TC), triglyceride (TG), low density lipoprotein cholesterol (LDLc), very low density lipoprotein cholesterol (VLDLc), high density lipoprotein cholesterol (HDLc) and the activities of catalase (CAT), glutathione peroxidase (GPx), glutathione-S-transferase (GST) as well as the levels of conjugated diene (CD) and thiobarbituric acid reactive substance (TBARS) in liver and kidney, were also shifted towards the control level in composite extract co-administered. Although extracts of the individual plants showed activity when administered separately, their effect was much more pronounced when they were administered together. This study highlighted the potentials of these herbs in management of type I diabetes. Similarly, studies conducted at St. John's Hospital, Bangalore suggested that *Coccinia cordifolia* extract has a potential hypoglycemic action in patients with mild diabetes (Kuriyan *et al.*, 2008).

Oral administration of *Coccinia indica* leaf extract (CLEt) (200 mg/kg body weight) for 45 days resulted in a significant reduction in thiobarbituric acid reactive substances and hydroperoxides in diabetic rats. The extract also causes a significant increase in reduced glutathione, superoxide dismutase, catalase, glutathione peroxidase and glutathione-S-transferase in liver and kidney of streptozotocin diabetic rats, which clearly shows the antioxidant property of Coccinia leaf extract (Venkateshwaran and Pari, 2003) and its usefulness in management of diabetes. The studies conducted by Baghel *et al.*, (2011) suggested that aqueous extract of *Coccinia indica* is toxicologically safe for oral administration.

Post- and pre-treatment anti-inflammatory activities of the aqueous extract of fresh leaves of *Coccinia indica* was studied in rats using the carrageenan-induced paw oedema method at various dose levels by Niazi *et al.*, (2009). In post-treatment studies, a dose-dependent anti-inflammatory effect was observed in the dose range of 25–300 mg/kg body. The effect was equivalent to diclofenac (20 mg/kg) at 50 mg/kg, but it was significantly pronounced at higher doses. Effectiveness of extract in the early phase of inflammation suggested the inhibition of histamine and serotonin release.

References

Baghel, S.S., Danai, S., Soni, P., Singh, P., andShivahare, Y. 2011. Acute toxicity study of aqueous extract of *coccinia indica* (roots). Asian J. Res. Pharm. Sci., (2011), 1(1): 23-25.

Bose, T.K. and Som, M.G. 1986. Vegetable crops in India. Naya Prokash, Calcutta.

Chandira, M., Vankateswarlu, B.S., Gangwar, R.K., Sampathkumar, K.P., Bhowmik, D., Jayakar, B., and Rao, C.V.2010. Studies on anti-stress and free radical scavenging activity of whole plant of *Coccinia Indica* Linn, Int RJ Pharm Sci., (2010), 01: 0050.

Gopalan, C., B.V. Rama Sastri and S.C. Balasubramanian. 1985. Nutritive Value of Indian Foods. National Institue of Nutrition, ICMR, Hyderabad, India.

Hossain, M.Z., Shibib, B.A., Rahman, R. 1992. Hypoglycemic effects of *Coccinia indica*: inhibition of key gluconeogenic enzyme, glucose-6-phosphatase. Indian J Exp Biol., (1992), 30(5):418-20.

Sakharkar, P., and Chauhan B. 2017. Antibacterial, antioxidant and cell proliferative properties of *Coccina grandis* fruits. Avicenna J Phytomed., (2010) 7(4): 295-307.

Kumar, P., Sudheesh, S., Viayalakhsmi, N. R. 1993. Hypoglycaemic Effect of *Coccinia indica*: Mechanism of Action. Planta Med., (1993), 59(4): 330-332.

Kuriyan, R., Rajendran, R., Bantwal, G., and Kurpad, A.V.2008. Effect of supplementation of *Coccinia cordifolia* extract on newly detected diabetic patients. Diabetes Care, (2008), 31(2):216-20.

Mallick, C., Chatterjee, K., Mandal, U., and Ghosh, D. 2007. Antihyperglycaemic, Antilipidperoxidative and Antioxidative Effects of Extracts of *Musa paradisiaca* and *Coccinia indica* in Streptozotocin-Induced Diabetic Rat. Ethiopian Pharmaceutical Journal, (2007), 25 (1): 9-22.

Niazi, J., Singh, P., Bansal, Y., and Goel, R K. 2009. Antiinflammatory, analgesic and antipyretic activity of aqueous extract of fresh leaves of *Coccinia indica*, Journal of Inflammopharmacology, (2010), 17:239-244.

Rastogi, R.P. 1998. Mehrotra BN Compendium of Indian medicinal plants, 1:115.

Ray, A.B. and Kundu, S. 1987. Chemical examination of Coccinia indica fruits. J of Indian Chem. Soc ., (1987), LXIV:776–777.

Shaheen, S. J., Bolla, K., Vasu, K., and Charya, M. A. S. 2009. Antimicrobial activity of the fruit extracts of *Coccinia indica* African J of Biotech., (2009), 24: 7073-7076.

Venkateswaran, S., and Pari, L. 2003. Effect of Coccinia indica leaves on antioxidant status in streptozotocin-induced diabetic rats. J Ethnopharmacol., (2003) 84(2-3):163-8.

Yadav, G., Mishra, A., and Tiwari, A. 2010. Medical properties of ivy gourd (Cephalandra indica): A Review. International Journal of Pharma. Research & Developmentonline; 2: 0974-9446.

2.6. Snake Gourd (*Trichosanthes cucumerina*)

Trichosanthes is one of the large genera having too many species of *which* 22 occur in India. The genus *Trichosanthes* is native to Southern and Eastern Asia, Australia and Islands of the western Pacific. Its centre of origin is believed to be India or Indo-Malayan region. Its reference in local languages is mostly found in Malay Peninsula, Moluccas and the Philippine islands. It was probably domesticated in ancient times in India. The common names for Trichosanthes in English are Chinese cucumber and snake gourd (derived from the long snake-like gourds produced by *Trichosanthes cucumerina*).

The regional names of snake gourd or snake tomato is as Chichinga/Chichinge in Bengali, potlakaaya in Telugu, pudalankaai in Tamil, aduvalakaayi in Kannada, padavalanga in Malayalam, galartori in Punjabi, padavali in Gujarathi, Chachinda in Hindi.

Cultivars/Varieties: The popular varieties of snake gourd under cultivation in India are CO-1, Kaumudi, Baby, TA-19 and Manusree.

Uses

The fruit is usually consumed as a vegetable due to its good nutritional value. The plant is rich in flavonoids, carotenoids and phenolic compounds. *Trichosanthes cucumerina* has a promising place in the Ayurvedic and Siddha system of medicine due to its various medicinal values like antidiabetic, hepatoprotective, cytotoxic, anti inflammatory, larvicidal effects.

Nutritional Components of Snake gourd

Constituents	Quantity	Constituents	Quantity
Moisture (g/100g)	94.6	Calcium (mg/100g)	26
Carbohydrates (g/100g)	3.3	Phosphorus (mg/100g)	20
Protein (g/100g)	0.5	Iron (mg/100g)	0.3
Fat (g/100g)	0.3	Thiamine (mg/100g)	0.04
Minerals (g/100g)	0.5	Riboflavin (mg/100g)	0.06
Fibre (g/100g)	0.8	Niacin (mg/100g)	0.3
Energy (Kcal)	18	Vitamin C (mg/100g)	2

(Gopalan *et al.*, 1985)

Phytochemicals

Tricosanthes cucumerina is a rich source of nutrition. Its nutrients constitute of carbohydrates, proteins, fat, fibre, minerals like calcium, phosphorus and iron, and vitamins like vitamin A, E, C and C complex. The total phenolics and flavonoids content is 46.8% and 78.0% respectively (Adebooye, 2008). The triterpenes found are 23, 24-dihydrocucurbitacin D, 23,24-dihydrocucurbitacin B, cucurbitacin B, 3β-hydroxyolean- 13(18)-en-28-oic acid, 3-oxo-olean-13(18)-en-30- oic acid and the sterol 3-*O*-β-D-glucopyranosyl-24ξ-ethylcholest- 7,22-dien-3β-ol (Jiratchariyakul and Frahm, 1992) besides sterols 2 β-sitosterol stigmasterol (Datta, 1987). Low amount of chemical substances like oxalate, phytates and tannins are also present. A galactose-specific lectin and ribosome-inactivating protein named trichoanguin13 are present in aerial parts (Chow *et al.*, 1999; Padma *et al.*, 1999). The bulk of carotenoids made of lutein is present in the concentration of 15.6–18.4 mg/100g FW (Anuradha and Bhide, 1999). The α-carotene contents were 10.3–10.7 mg/100g FW and the β-carotene contents were found to be 2.4–2.8 mg/100g. It also has lycopene content of around 16.0 and 18.1 mg/100g FW.

Health Benefits of Snake gourd based on Traditional Knowledge/Folk-lore practices

T. cucumerina L. var. *cucumerina* is Ayurvedic medicinal plant belonging to the family Cucurbitaceae. More than 16 herbal products are available in the market in which *T. cucumerina* L. var. *cucumerina* is one of the important ingredients (Kage *et al.*, 2008).

• The plant is effectively used in various indications, whole plant extract as depurative stomachic in nineteenth century Cayenne, French Guiana (Heckel, 1897), to reduce congestion on con-gestive cardiac failure (Pullaih, 2006)

• Aphrodisiac (Sood *et al.*, 2005).

• Leaf of snake gourd is used for inter-mittent fever (Kirtikar and Basu, 2000), skin disease (Chopra *et al.*, 1969), bald patches of alopecia (Anonymous, 1976).

• Fruit is regarded as anthelmintic, vomitive (Jeffrey, 1984), antidia-betic (Devendra, 2010), for boil (Srivstava *et al.*, 2003).

• Seeds are considered anthelmintic, and anti fibrile (Nadakarni, 2005).

• Various species of *Trichosanthes* has been in Chinese and Indian medicinal use. According to Oriental Materia Medica (Hong *et al.*, 1986)

• In China, roots are used for diabetes, skin swellings like boils and furuncles. Fresh root has anti-convulsant activity. Bulbous part of the root is used as a hydragogue and cathartic. Root is abortifacient, alexiteric, anthelmintic, anti-septic, astringent, diuretic and emetic.

• The pericarp of the fruit is especially used for relieving stagnancy of qi circulation in the chest,

• Bothe the fruit and seed is especially useful as a moistening agent for treating constipation.

• The root as a medicine was first officially described more than 800 years ago (in 1062 AD in fact by *Tujing Bencao*). *Trichosanthes* root is described as sweet, slightly bitter, and slightly cold, is said to be particularly useful for treating diabetes and skin swellings (it is known to decrease pus formation and thus treat furuncles, carbuncles, boils, and abscesses).

• The stalk and leaves of the plant are also used as a folk remedy for feverish diseases as reported by Shen Ruoxing (2001).

Among the Tribals in India, *Tricosanthes cucumerina* is used in the treatment of

- Headache, alopecia, fever, abdominal tumors, bilious, boils, acute colic, diarrohea, haematuria and skin allergy.
- It is used as an abortifacient, vermifuge, stomachic, refrigerant, purgative, malaria, laxative, hydragogue, hemagglutinant, emetic, cathartic, bronchitis, a nd anthelmintic (Nadakarni, 2005).
- Root is used as purgative and tonic. Two ounces of root juice has a drastic purgative action. Roots are used for expelling worms.
- Leaf juice is rubbed over the whole body in remittent fevers. Dried leaf has anti-spasmodic property.
- An infusion of tender shoots and dried capsules is aperient, and the expressed juice of the leaves is emetic.
- The leaves and stems are used for bilious disorders and skin diseases and as an emmenagogue. Leaf is alexiteric, astringent, diuretic and emetic.
- In Ayurveda and Indian Materia Medica, the fruit is considered to be anthelmintic. The dried capsules are given in infusion or in decoction with sugar to assist digestion. The seed is said to be cooling. The dried seeds are used for its anthelmintic and anti-diarrhoeal properties. Seeds have anti-bacterial, anti-spasmodic, antiperiodic and insecticidal properties. It is used as abortifacient, acrid, aphrodisiac, astringent, bitter, febrifuge, purgative, toxic, trichogenous (Nadakarni, 2005; Madhava *et al.*, 2008).

Therapeutic Properties of Snake Gourd

On administration of hot aqueous extract of root tubers of *Trichosanthes cucumerina* against carrageenin induced mouse's hind paw oedema, it exhibited significant anti-inflammatory activity (Kolte *et al.*, 1997). Kongtun *et al.*, (1999) tested the root extract and fruit juice of *Trichosanthes cucumerina* L. for its cytotoxicity effect against four human breast cancer cell lines and lung cancer cell lines and one colon cancer cell line. The root extract inhibited the cancer cell growth more strongly than the fruit juice. Similarly, Kar *et al.*, (2003) investigated the hypoglycaemic activity of snake gourd with crude ethanolic extract of *Tricosanthes cucumerina* which showed significant blood glucose lowering activity in alloxan diabetic albino rats. Hot water extract of aerial parts of snake gourd showed improved glucose tolerance and tissue glycogen in non insulin dependent diabetes mellitus induced rats indicating its anti-diabetic activity (Arawwawala *et al.*, 2009).

Sathesh *et al.*, (2009) found that the methanolic extract of the whole plant of *Tricosanthes cucumerina* showed good hepatoprotective activity against carbon tetrachloride induced heapatotoxicity. Acetone extract of leaves of Tricosanthes *cucumerina* showed moderate larvicidal effects (Rahuman *et al.*,2008). Ethanol extract of whole plant of *Trichosanthes cucumerina* L. var. *cucumerina* exhibited antiovulatory activity in female albino rats (Kage *et al.*, 2009).

Arawwawala *et al.*, (2009) with hot water extract of *Trichosanthes cucumerina*, showed a significant protection against ethanol or indomethacin induced gastric damage increasing the protective mucus layer, decreasing the acidity of the gastric juice and antihistamine activity. Dose dependent gastroprotective effects were observed in the alcohol model in terms of the length and number of gastric lesions mediated by alcohol in wistar stain rats (Sandhya *et al.*, 2010). Though all these findings are limited laboratory trials, it still supports the claims to an extent. The traditional practices followed were mostly time tested based on practice over several decades on human subjects, as the modern animal model facilities were not available earlier times. Even those claims stand the scrutiny in the light of present scientific knowledge. Extensive clinical studies with in light of emerging cutting-edge technologies may further help to probe into the depth of knowledge.

References

Adebooye, O.C. 2008. Phytoconstituents and anti-oxidant activity of the pulp of snake tomato (*Tricosanthes cucumerina*), African Journal of Traditional, Complimentary and Alternative Medicines, (2008), 5 (2): 173-179.

Anonymous. 1976. Welth of India. Vol-10, CSIR, New Delhi, pp-291.

Anuradha, P., Bhide, S.V. 1999. An isolectin complex from *Trichosanthes anguina* seeds, Phytochemistry, (1999), 52(5): 751-8.

Arawwawala, L.D., Thabrew, M.I., Arambewela, L.S. 2010. Gastroprotective activity of *Trichosanthes cucumerina* in rats, J Ethnopharmacology, (2010), 127(3):750-754. Doi: 10.1016/j.jep.2009

Arawwawala, M., Thabrew, I., and Arambewela, L., 2009. Antidiabetic activity of *Trichosanthes cucumerina* in normal and streptozotocin–induced diabetic rats; International Journal of Biological and Chemical Sciences, (2009), 3(2): 56.

Azeez, M.A., Morakinyo, J.A. 2004. Electrophoretic characterization of crude leaf proteins in *Lycopersicon* and *Trichosanthes* cultivars, African Journal of Biotechnology, (2004), 3(11):585-587.

Bose, T.K. and Som, M.G.1986. Vegetable crops in India. Naya Prakash, Calcutta.

Chopra, R. N., Nayar, S. L., and Chopra, I. C 1969. Glossary of Indian Medicinal Plants. Counsil of scientific and industrial research, New Delhi, pp-248.

Choudhury, B. 1967. Vegetables. National Book Trust, New Delhi, India, 1967, 214.

Chow, L.P., Chou, M.H., Ho, C.Y., Chuang, C.C., Pan, F.M., Wu, S.H., Lin, J.Y. 1999. Purification, characterization and molecular cloning of trichoanguin, a novel type I ribosome inactivating protein from the seeds of *Trichosanthes anguina*, Biochemical Journal, (199), 338: 211–219.

Datta, S.K. 1987. Fatty acid composition in developing seeds of *Trichosanthes cucumerina* L, Biological Memoirs, (1987), 13(1): 69-72.

Gildemacher, B.H., Jansen, G.J., Chayamarit, K. 993. Plant Resources of South-East Asia No 8. Vegetables, Trichosanthes L. In: Siemonsma JS, Kasem P, Pudoc Scientific Publishers, Wageningen, Netherlands.1993, 271-274.

Gopalan, C., B.V.Rama Sastri and S.C.Balasubramanian. 1985. Nutritive Value of Indian Foods. National Institue of Nutrition, ICMR, Hyderabad, India.

Heckel, E 1897. Les Plantes Médicinales et Toxiques de la Guyane Francaise. 93 pp. Macon, France: Protat Freres.

Hong-Yen Hsu, .1986. Oriental Materia Medica: A Concise Guide, 1986 Oriental Healing Arts Institute, Long Beach, CA.

Indian medicinal plants compendium of 500 species, Orient Longman Pvt. Ltd. Chennai, 2002, 320-322.

Jeffrey, C. 1984. Cucurbitaceae, pp. 457-518. In: Stoffers, A.L. and J.C. Lindeman, eds., Flora of Surinam. Vol. 5, Part 1. Leiden: E.J. Brill.

Jiratchariyakul, W., and Frahm, A.W. 1992. Cucurbitacin B and dihydrocucurbilacin B from Trichosanthes cucumerina, J. Pharm. Sci., (1992), 19 (5):12-15.

Kage, D.N., Rajanna, L., Sheetal, C., and Seetharam, Y.N., 2008. In vitro Clonal Propagtion of Trichosanthes cucumerina L. var. cucumerina. Plant Tissue Cult. & Biotech., (2008), 18(2): 103-111.

Kage, D.N., Vijay, K.B., and Mala, S. 2009. Effect of ethanol extract of whole plant of Tricosanthes cucumerina Var.Cucumerina on gonadotropins, ovarian follicular kinetics and estrous cycle for screening of anti fertility activity in albino rats, Int. J. Morphol., (2009), 27(1):173-182.

Kage, D.N., Vijaykumar, B. M., and Seetharam, Y. N. 2010. Folklore Medicinal Plants of Gulbarga District, Karnataka, India. E-Journal of Indian Medicine., (2010), 3(1): 23-30.

Kar, A., Choudhury, B.K., and Bandyopadhyay, N.G. 2003. Comparative evaluation of hypoglycaemic activity of some Indian plants in alloxan diabetis rats, Journal of Ethnopharmacology, (2003), 84(1):105-108.

Kenoth, R., Komath, S.S., and Swamy, M.J. 2003. Physicochemical and saccharide-binding studies on the galactosespecific seed lectin from Trichosanthes cucumerina, Arch Biochem Biophys, (2003), 413 (1): 131-8.

Kirana, H., and Srinivasan, B. 2008.Tricosanthes cucumerina improves glucose tolerance and tissue glycogen in noninsulin dependant diabetis mellitus induced rats, Indian Journal of Pharmacology, (2008), 345-8.

Kirtikar and Basu. 2000. Indian Medicinal Plants. Sri Satguru Publications, New Delhi. (5):1545-1547.

Kolte, R.M., Bisan, V.V., Jangde, C.R., Bhalerao, A.A. 1997. Anti-inflammatory activity of root tubers of Trichosanthes cucumerina (LINN) in mouse's hind paw oedema induced by carrageenin, Indian Journal of Indigeneous Medicines, (1997), 18(2) :117- 21.

Kongtun, S., Jiratchariyakul, W., Mongkarndi, P., Theppeang, K., Sethajintanin, I., Jaridasem, S., Frahm, A.W. 1999. Cytotoxic properties of root extract and fruit juice of Tricosanthes cucumerina, J Phytopharm., (1999),6 (2): 1-9.

Madhava, K.C., Sivaji, K., Tulasi, K.R. 2008. Flowering Plants of Chitoor Dist A.P. India, Students Offset Printers, Tirupati, 2008, 141.

Nadakarni, A. K. 2005. Indian Materia Medica. Vol-I, Popular Prakashan, Mumbai, pp-1235-1236.

Padma, P., Komath, S.S., *et al.*, 1999. Purification in high yield and characterization of a new galactosespecific lectin from the seeds of *Trichosanthes cucumerina*. Phytochemistry Oxford, (1999), 50(3): 363-371.

Pullaih, T. 2006. Encyclopaedia of World Medicinal Plants. Vol-IV, Regency Publication New Delhi. Pp-1977.

Rahuman, A.A., ad Venkatesan, P. 2008. Larvicidal efficacy of five cucurbitaceous plant leaf extracts against mosquito species, Parasitol Res, (2008),103(1): 133-9.

Sandhya, S., Vinod, K.R., Chandra Sekhar, J., Aradhana, R., Nath, V. S. 2010. An updated review on *Tricosanthes Cucumerina* L. International Journal of Pharmaceutical Sciences Review and Research, (2010), 1(2): 56-60.

Sathesh, K.S., Ravi, K.B., and Krishna, M.G. 2009. Hepatoprotective effect of *Tricosanthes cucumerina* L on carbon tetra chloride induced liver damage in rats. J Ethnopharmacol., (2009), 123(2):347-50.

Shen Ruoxing. 2001. Distinguishing the uses of related medicinals, RCHM News; Spring 2001, 13-15.

Sood, S. K., Sarita, R., and Lakhanpal, T. N. 2005. Ethnic Aphrodisiac plants. Scientific publishers (India) Jodhpur, pp-118.

Srivastava, R. C., Singh, V. P., and Singh, M. K. 2003. Medicinal plants of Jaunapur District U. P. India. In Singh *et al.*, (ed.) Ethnobotany and Medicinal Plants of India and Nepal. Vol-I, Scientific publishers (India).

2.7. Ridge and Sponge Gourd (*Luffa*)

Luffa is a genus of tropical and subtropical vines classified in the cucumber family Cucurbitaceae. *Luffa* consists of two cultivated species and two wild species. The two cultivated species are *Luffa accutangula* (Ridge or Ribbed gourd) and *Luffa cylindrica* (Sponge gourd). In everyday non-technical usage, the name, also spelled **loofah**, usually refers to the fruit of the two species *Luffa aegyptiaca* and *Luffa acutangula*. The wild ones are *Luffa graveolens* and *Luffa aphanin*. The fruit of these two species (L. *aegytiaca* and L. *acutangula*) is cultivated and eaten as a vegetable. This vegetable is popular in China and southeast Asia. When the fruit is fully ripened it is very fibrous. The fully developed fruit is the source of the loofah scrubbing sponge which is used in bathrooms and kitchens as a sponge tool. Luffa are best eaten when small (less than 12 cm) and still green. The ridge gourd has prominent ridges on the surface and spong gourd has a smooth surface. They are believed to be native to Asia and Africa and its Sanskrit name is 'Koshataki' signifying its existence in India for a very long time.

Cultivars/Varieties: The popular varieties of ridge gourd are Pusa Nasdar, CO-1, CO-2, Haritham, Deepthi, Desi Chaitali, Kankan Harita and Phule Sucheta. There is a cultivar in Bihar called 'Satputia' which is hermaphrodite and produces smaller fruits in cluster. In sponge gourd Pusa Chikni is very popularly grown and preferred in North India.

Uses

Largely in India and east Asian countries it is used for vegetable. Small quantities are also used for making body sponge (body scrubber). The widely consumed one is called Jhinga or ridge gourd (Luffa *aphanina*). The fruits are eaten as vegetable. Fruit is demulcent, diuretic and nutritive. The seeds possess purgative and emetic properties. The pounded leaves are applied locally to splenitis, haemorrhoides and leprosy. The juice of the fresh leaves is dropped into the eyes of children in granular conjunctivitis and also to prevent the lids adhering at night from excessive meibomian secretion (Vashista, 1974).

Nutritional Composition of Ridge Gourd

Constituents	Quantity	Constituents	Quantity
Moisture (g/100g)	95.2	Calcium (mg/100g)	18
Carbohydrates (g/100g)	3.4	Phosphorus (mg/100g)	26
Protein (g/100g)	0.5	Iron (mg/100g)	0.5
Fat (g/100g)	0.1	Thiamine (mg/100g)	--
Minerals (g/100g)	0.5	Riboflavin (mg/100g)	0.1
Fibre (g/100g)	0.5	Niacin (mg/100g)	0.2
Energy (Kcal)	17	Vitamin C (mg/100g)	5

(Gopalan *et al.*, 1985)

Phytochemicals

The fruits contain a bitter principle, luffeine. The seeds contain glycerides of palmitic, stearic and myristic acids. Luffaculin (isolated from seeds of *Luffa aphanina*) and luffin-a and luffin-b isolated from seeds of *Luffa cylindrical* are the other phytochemicals reported in Luffa species. *Luffa egyptiaca* is used in traditional medicine to treat snakebite as it has inhibitory activities on *Naja nigricolis* venom protease (Ibrahim *et al.*, 2011). Studies have shown seven oleanane-type triterpene saponins, acutosides A-G in Luffa.

Health Benefits of Luffa based on Traditional Knowledge/Folklore Practices

- Decoction of leaves is used for amenorrhea.
- Poultice of leaves is used for hemorrhoids. Juice of fresh leaves for conjunctivitis, and also externally for sores and various animal bites. Leaf sap used as eyewash in conjunctivitis in West Africa and sores caused by guinea works. In Mauritius too, seeds are eaten to expel intestinal worms.
- Both seed oil and leaf juice are is used for dermatitis and eczema. The infusion of seeds is used as purgative and emetic.
- Roots is used for dropsy and as laxative.
- Leaf and fruit juice are used to treat jaundice in India. In Java, leaf decoction is used for uremia and amenorrhea.
- In Bangladesh, pounded leaves is used for hemorrhoids, splenitis, leprosy.
- Fruits and seeds used in herbal preparations for treatment of venereal diseases.

Therapeutic Properties of Ridge Gourd

Ethyl acetate fractions partitioned from ethanol extracts of *Luffa egyptiaca* was found to completely inhibit the *N. nigricolis* venom protease activity (Ibrahim *et al.*, 2011). Hepatoprotective activity of hydroalcoholic extract of *Luffa aphanina* (HAELA) was studied which demonstrated that endogenous antioxidants and inhibition of lipid peroxidation of membrane contribute to hepatoprotective activity of HAELA (Jadav *et al.*, 2010). Luffin P1, the smallest ribosome-inactivating peptide from the seeds of *Luffa cylindrica* was found to have anti-HIV-1 activity in HIV-1 infected C8166 T-cell lines and be able to bind with HIV Rev Response Element (Ng *et al.*, 2011). High-performance liquid chromatography analysis showed that a total of eight compounds in the dried gourds without skin was about 1% and the results demonstrated that the consumption of sponge gourds can supply some antioxidant constituents to human body (Du *et al.*, 2006). Two Triterpenoids (sapogenins 1 and 2) isolated *from Luffa cylindrica* were subjected to immunomodulatory activity studies in male Balb/c mice. Sapogenins 1 and 2 showed significant dose-dependent decrease and increase in lymphocyte proliferation assay and phagocytic activity of macrophages suggesting its immunostimulatory effect in mice (Khajuria *et al.*, 2007). A dose dependent larvicidal activity of seven cucurbits including *Luffa aphanina* was demonstrated by Prabhakar and Jabanesan (2007).

References

Bose, T.K. and Som, M.G.1986. Vegetable crops in India. Naya Prokash, Calcutta.

Du, Q., Xu, Y., Li, L., Zhao, Y., Jerz, G., and Winterhalter, P. 2006. Antioxidant constituents in the fruits of *Luffa cylindrica* (L.) Roem. J Agric Food Chem., (2006), 54(12):4186-90.

Gopalan, C., B.V. Rama Sastri and S.C. Balasubramanian. 1985. Nutritive Value of Indian Foods. National Institue of Nutrition, ICMR, Hyderabad, India.

Gopalan, C., Sastri, B.V.R., and Balasubramanian, S.C. 1985. Nutritive Value of Indian Foods. National Institue of Nutrition, ICMR, Hyderabad, India.

Ibrahim, M.A., Aliyu, A.B., Abusufiyanu, A., Bashir, M., Sallau, A.B. 2011. Inhibition of *Naja nigricolis* (Reinhardt) venom protease activity by *Luffa egyptiaca* (Mill) and Nicotiana rustica (Linn) extracts. Indian J Exp Biol., (2011), 49(7):552-4.

Jadhav, V.B., Thakare, V.N., Suralkar, A.A., Deshpande, A.D., and Naik, S.R. 2010. Hepatoprotective activity of Luffa aphanina against CCl4 and rifampicin induced liver toxicity in rats: a biochemical and histopathological evaluation. Indian J Exp Biol., (2010), 48(8):822-9.

Khajuria, A., Gupta, A., Garai, S., and Wakhloo, B.P. 2007. Immunomodulatory effects of two sapogenins 1 and 2 isolated from *Luffa cylindrica* in Balb/C mice. Bioorg Med Chem Lett., (2007), 17(6):1608-12.

Ng, Y.M., Yang, Y., Sze, K.H., Zhang, X., Zheng, Y.T., and Shaw PC.2011. Structural characterization and anti-HIV-1 activities of arginine/glutamate-rich polypeptide Luffin P1 from the seeds of sponge gourd (*Luffa cylindrica*). J Struct Biol., (2011), 174(1):164-72.

Prabakar, K., and Jebanesan, A. 2004. Larvicidal efficacy of some Cucurbitacious plant leaf extracts against *Culex quinquefasciatus* (Say). Bioresour Technol., (2004), 95(1):113-4.

Vashista, P.C. 1974. Taxonomy of Angiosperms. P.B.M. Press, New Delhi, India.

2.8. Cucumber (*Cucumis sativus*, L.)

Cucumber belongs to the *Cucumis* genus which comprises about 30 species distributed over two distinct geographic areas i) south east of Himalayas is an important region of Asiatic group ii) and the other African group comprising large part of Africa, Middle East, Central Asia extending to Pakistan and South Arabia. Cucumber and Muskmelon are the two important crops which belongs to this genus. The cucumber is believed to be originated in India but is now cultivated all over the world. References indicate that cucumber has been cultivated in India for over 3000 years. This vegetable is one of the oldest cultivated crops and believed to be originating in the northern plains of India. Armenian cucumbers (*Cucumis melo* var. *flexuosus*) are long, crispy, and thin-ribbed, curved, and have light green color. Although they are grouped botanically in melons family they look and taste like cucumbers. Small size varieties such as gherkins, *American dills* and French-cornichons are very small in size and usually preferred in pickling.

Varieties: The varieties of cucumbers are classified into three main varieties: "slicing", "pickling", and "burpless". Poinsette, Japanese Long green, Straight Eight, Sheetal, Pusa Sanjog, Priya, Swarna Poorna, DVRM-1, IIHR-177-1 are some of the varieties widely cultivated in India.

Uses

In India Cucumbers are popularly known as 'khira' and gherkins are part of the cucurbits variety of crops extensively grown in tropics, subtropics and milder temperate zones of India, mainly as a salad crop. It is eaten raw, pickled or cooked. A number of varieties like Poona khira (a small-size pale-green fruit cultivated in western Maharashtra), Balam khira (cultivated in Saharanpur in Uttar Pradesh); and the Darjeeling and Sikkim varieties (grown in hills of North Bengal) have become very popular. Cucumbers are used in different ways. Fresh, clean cucumbers may be enjoyed as they are without any additions. Its cubes are a great addition to vegetable/fruit salads. It is also being used in some variety of curry preparation in south India with buttermilk and yogurt. Finely chopped fresh slices mixed with yogurt, cumin, coriander, pepper, and salt to make Indian *cucumber raita*. Cucumber juice is a very good healthy drink. Fine slices also added in delicious Spanish cold tomato and cucumber soup, **gazpacho**. Its rind is also used in the preparation of pickles.

Nutritional aphanina of Cucumber

Constituents	Quantity (per 100g)	Constituents	Quantity (per 100g)
Moisture	96.3 g	Calcium	10 mg
Carbohydrates	2.5 g	Phosphorus	25 mg
Protein	0.4 g	Iron	1.5 mg
Fat	0.1 g	Thiamine	0.03 mg
Minerals	0.3 g	Riboflavin	
Fibre	0.4 g	Niacin	0.2 mg
Energy (Kcal)	13	Vitamin C	7 mg

(Gopalan *et al.*, 1985)

Phytochemicals

Cucumbers are now known to contain lariciresinol, pinoresinol, and secoisolariciresinol-three lignans that have a strong history of research in connection with reduced risk of cardiovascular disease as well as several cancer types, including breast, uterine, ovarian, and prostate cancers. Fresh extracts from cucumbers have recently been shown to have both antioxidant and anti-inflammatory properties. Substances in fresh cucumber extracts help scavenge free radicals, help improve antioxidant status, inhibit the activity of pro-inflammatory enzymes like cyclo-oxygenase 2 (COX-2), and prevent over production of nitric oxide in situations where it could pose health risks. At the top of the phytonutrient list for cucumbers are its cucurbitacins, lignans, and flavonoids. These three types of phytonutrients found in cucumbers provide us with valuable antioxidant, anti-inflammatory, and anti-cancer benefits. Specific phytonutrients provided by cucumbers include flavonoids like apigenin, luleolin, quercetin and kaempferol. Lignans like pinoresinol, lariciresinol and secoisolariciresinol. Triterpenes like cucurbitacin A, cucurbitacin B, cucurbitacin C and cucurbitacin D.

They are also a very good source of the enzyme-cofactor molybdenum. They are also a good source of free radical-scavenging vitamin C; heart-healthy potassium and magnesium, bone-building manganese, and energy-producing vitamin B5. They also contain the important nail health-promoting mineral silica.

Health Benefits of Cucumbers Based on Traditional Knowledge/Folklore Practices

Cucumbrs are found to have many curative properties for common ailments. Because of its fibre content, it removes human constipation and good for digestion. The fruits are much used during summer as a cooling food. They are used as salads and for cooking curries. The tender fruits are preferred for pickling, kernels of the seeds are used in confectionary (Chakravarty, 1982). Fruit is demulcent. Seeds are cooling, tonic, diuretic and anthelmintic. Leaves are along with cumin seeds administered in throat affections (Vashista, 1974).

- It is one of the very low calories vegetable; provides just 15 calories per 100 g. It contains no saturated fats or cholesterol.
- Cucumber peel is a good source of dietary fiber that helps reduce constipation and offers some protection against colon cancers by eliminating toxic compounds from the gut.
- It is a very good source of potassium, an important intracellular electrolyte. Potassium is a heart friendly electrolyte and helps reduce blood pressure and heart rates by countering effects of sodium.
- Cucumbers have mild diuretic property probably due to their high water and potassium content, which helps in checking weight gain and high blood pressure.
- Raw cucumber, when applied on the skin, can help reduce heat and inflammation.
- The diuretic, cooling and cleansing property of cucumber makes it good for skin.

- Fresh cucumber juice can provide relief from heartburn, acid stomach, gastritis and even ulcer.
- Placing a cucumber slice over the eyes not only soothes them, but also reduces swelling.
- Daily consumption of cucumber juice helps control cases of eczema, arthritis and gout.
- Cucumber has been found to be beneficial for those suffering from lung, stomach and chest problems.
- Cucumber contains Erepsin, the enzyme that helps in protein digestion.
- Cucumber juice is said to promote hair growth, especially when it is added to the juice of carrot, lettuce and spinach.
- Cucumber juice, when mixed with carrot juice, is said to be good for rheumatic conditions caused by excessive uric acid in the body.
- Cucumber can prove to be beneficial for those suffering from diseases of the teeth and gums, especially in cases of pyorrhea.
- Being rich in minerals, cucumber helps prevent splitting of nails of the fingers and toes.
- Cucumber has been associated with healing properties in relation to diseases of the kidney, urinary bladder, liver and pancreas.
- Those suffering from diabetes have been found to benefit from the consumption of cucumber/cucumber juice.

Therapeutic Properties of Cucumber

In animal studies, fresh extracts from cucumber have been shown to provide specific antioxidant benefits, including increased scavenging of free radicals and increased overall antioxidant capacity. Fresh cucumber extracts have also been shown to reduce unwanted inflammation in animal studies. Cucumber accomplishes this task by inhibiting activity of pro-inflammatory enzymes like cyclo-oxygenase 2 (COX-2), and by preventing overproduction of nitric oxide in situations where it could increase the likelihood of excessive inflammation.

Research on the anti-cancer benefits of cucumber is still in its preliminary stage and has been restricted thus far to lab and animal studies. Interestingly, however, many pharmaceutical companies are actively studying one group of compounds found in cucumber–called cucurbitacins–in the hope that their research may lead to development of new anti-cancer drugs. Cucurbitacins belong to a large family of phytonutrients called triterpenes. Researchers have determined that several different signaling pathways (for example, the JAK-STAT and MAPK pathways) required for cancer cell development and cancer cell survival can be blocked by activity of cucurbitacins.

A second group of cucumber phytonutrients known to provide anti-cancer benefits are its lignans. The lignans – pinoresinol, lariciresinol, and secoisolariciresinol have all been identified within cucumber. Interestingly, the role of these plant lignans in cancer protection involves the role of bacteria in our digestive tract. When we consume plant lignans like those found in cucumber, bacteria in our digestive tract take hold of these lignans and convert them into enterolignans like enterodiol and enterolactone. Enterolignans can bind onto estrogen receptors and can have both pro-estrogenic and anti-estrogenic effects. Reduced risk of estrogen-related cancers, including cancers of the breast, ovary, uterus, and prostate has been associated with intake of dietary lignans from plant foods like cucumber.

References

Abiodun, O.A. 2010. Comparative Studies on Nutritional Composition of Four Melon Seeds Varieties. Pakistan Journal of Nutrition, (2010), 9(9): 905-908.

Bose, T.K. and Som, M.G.1986. Vegetable crops in India. Naya Prokash, Calcutta.

Chakravarty, H.L.1982. Fascicles of Flora of India. Botanical Survey of India, Calcutta.

Ghebretinsae, A.G., Thulin, M., and Barber, J.C. 2007. Relationships of cucumbers and melons unraveled: molecular phylogenetics of Cucumis and related genera (Benincaseae, Cucurbitaceae). Am J Bot., (2007), 94(7):1256-66.

Gopalan, C., Sastri, B.V.R., and Balasubramanian, S.C. 1985. Nutritive Value of Indian Foods. National Institue of Nutrition, ICMR, Hyderabad, India.

Hong, S.H., Choi, S.A., Yoon, H., et al., 2011. Screening of Cucumis sativus as a new arsenic-accumulating plant and its arsenic accumulation in hydroponic culture. Environ Geochem Health, (2011), 33(Suppl 1):143-149.

Kumar, D., Kumar, S., Singh, J., et al., 2010. Free radical scavenging and analgesic activities of Cucumis sativus l. fruit extract. J Young Pharm., (2010), 2(4):365-8.

Lee, D.H., Iwanski, G.B., and Thoennissen, N.H. 2010. Cucurbitacin: ancient compound shedding new light on cancer treatment. Scientific World Journal, (2010), 10:413-8.

Martinez, L., Thornsbury, S., and Nagai, T. 2006. National and international factors in pickle markets. Agricultural Economics Reports, No, 628, October 2006. Department of Agricultural Economics, Michigan State University, East Lansing, MI. 2006.

Milder, I.E.J., Arts, I.C.W., van de Putte, B., et al., 2005. Lignan contents of Dutch plant foods: a database including lariciresinol, pinoresinol, secoisolariciresinol and matairesinol. Br J Nutr., (2005), 93:393-402.

Nema, N, K., Maity, N., Sarkar, B., et al., 2011. Cucumis sativus fruit-potential antioxidant, anti-hyaluronidase, and anti-elastase agent. Arch Dermatol Res., (2011), 303(4):247-52.

Rios, J.L., Recio, M.C., Escandell, J.M., et al., 2009. Inhibition of transcription factors by plant-derived compounds and their implications in inflammation and cancer. Curr Pharm Des., (2009),15(11):1212-37.

Rios, J.L. 2010. Effects of triterpenes on the immune system. J Ethnopharmacol., (2010), 128(1):1-14.

Schrader, W.L., Aguiar, J.L., and Mayberry, K.S. 2002. Cucumber Production in California. Publication 8050. (2002). University of California Agricultural and Natural Resources, Davis, CA. 2002.

Sebastian, P., Schaefer, H., Telford, IR., et al., 2010. Cucumber (Cucumis sativus) and melon (C. melo) have numerous wild relatives in Asia and Australia, and the sister species of melon is from Australia. Proc Natl Acad Sci U S A. (2010), 107(32):14269-73.

Tang, J., Meng, X., Liu, H., *et al.*, 2010. Antimicrobial activity of sphingolipids isolated from the stems of cucumber (*Cucumis sativus* L.). Molecules, (2010), 15(12):9288-97.

Thoennissen, N.H., Iwanski, G.B. and Doan, N.B. 2009. Cucurbitacin B induces apoptosis by inhibition of the JAK/STAT pathway and potentiates antiproliferative effects of gemcitabine on pancreatic cancer cells. Cancer Res., (2009), 69(14):5876.

Vashista, P.C. 1974. Taxonomy of Angiosperms. P.B.M. Press, New Delhi, India.

2.9. Pumpkin, Squash, Marrow (Cucurbita sp)

Pumpkin, *Cucurbita moschata*, belongs to the family Cucurbitaceae. *Cucurbita* genus consists of about 27 species concentrated in the tropical regions of Central and South America comprising both wild and cultivated species. Basically, they are native to North America. They typically have a thick, orange or yellow shell, creased from the stem to the bottom, containing the seeds and pulp. Among these 27 species five viz. *C. moschata, C. maxima, C. pepo* (in America) *C. ficifolia* and *C. mixta* (in India) are widely cultivated world over. In India, pumpkin (*C. moschata*) is also called 'Kaddu', 'Kashiphal or Seethaphal'. It is widely cultivated in different parts of the country. The maximum area is in Uttar Pradesh, Karnataka, Tamil Nadu and Rajasthan.

Cultivars/Varieties: The popular varieties are Arka Suryamukhi, Arka Chandan, CO-1, CO-2, Early Yellow Prolific, and Pusa Alankar.

Uses

Pumpkins are used in different ways. Though, basically it is used as vegetables in India, in other countries, it is used to make soups, desserts and bread. In America pumpkin pie is made during the Thanksgiving dinner/meal. The pulp being yellow in color it is extensively used for blending with tomato pulp for making sauce. In dehydration powder form, it is used as thickening agent in several dishes. Being bland or slightly sweet, it is used for making desserts in some places. The seeds of pumpkin are extremely nutritious, and the roasted seeds are consumed as a healthy snack.

Nutritional Components in Pumpkin

Constituents	Quantity (per 100g)	Constituents	Quantity (per 100g)
Moisture	92.6 g	Zinc	0.32 mg
Carbohydrates	6.5 g	Magnesium	12 mg
Protein	1.0 g	Potassium	340 mg
Fat	0.1 g	Vitamin A	7384 µg
Minerals	0.6 g	Niacin	0.6 mg
Fibre	0.5 g	Vitamin C	9 mg
Energy (Kcal)	13 Kcal	Pantothenic acid	0.298 mg
Calcium	21 mg	Thiamine	0.05 mg
Phosphorus)	44 mg	Riboflavin	0.110 mg
Iron	0.8 mg	Vitamin E	1.06 mg

Source: USDA Nutrient Database

Pumpkin is incredibly rich in vital antioxidants and vitamins. It is very low in calories yet good source of vitamin A, flavonoid polyphenolic antioxidants like leutin, xanthins and carotenes. Pumpkin is believed to be a folk remedy for prostate problems, and is claimed to combat benign prostatic hyperplasia. Pumpkin seed oil contains essential fatty acids that help maintain healthy blood vessels, nerves and tissues. The medicinal properties of pumpkin include anti-diabetic, antioxidant, anti-carcinogenic, and anti-inflammatory (Yadav *et al.*, 2010).

Phytochemicals

Pumpkin is a store house of many antioxidant vitamins such as vitamin-A, vitamin-C, vitamin-E and dietary fibre. It is also an excellent source of many natural polyphenolic flavonoid compounds such as α and ß carotenes, cryptoxanthin, lutein and zeaxanthin. Carotenes convert into vitamin A inside the body. It is rich in B-complex group of vitamins like folates, niacin, vitamin B-6 (pyridoxine), thiamin and pantothenic acid. They comprise of L-tryptophan, a compound that has been found to be effective against depression. Several researchers have isolated/identified large number of phytochemicals from either fruit, seed or leaves of pumpkin, each having some or the other therapeutic value.

A novel ribosome-inactivating protein called 'Moschatin' from the mature seeds of pumpkin (*Cucurbita moschata*) is found to have large medicinal value and has been purified by Heng *et al.*, (2003). Cucurmosin is a novel type 1 ribosome-inactivating protein (RIP) isolated from the sarcocarp of *Cucurbita moschata* (pumpkin) (Zhang *et al.*, 2011). An antifungal peptide, Cucurmoschin, was isolated from the seeds of the black pumpkin by Wang and Ng (2003). Jiang (2011), obtained two new tetrasaccharide glyceroglycolipids from pumpkin that is novel in itself.

Li *et al.*, (2009) also isolated two new phenolic glycosides from the seeds of *Cucurbita moschata*. Their structures were elucidated as (2-hydroxy)phenylcarbinyl 5-O-benzoyl-beta-D-apiofuranosyl(1à2)-beta-D-glucopyranoside (1) and 4-beta-D-(glucopyranosyl hydroxymethyl) phenyl5-O-benzoyl-beta-D-apiofuranosyl(1à2)-beta- D-gluco-pyranoside (2) on the basis of spectroscopic analysis and chemical evidence.

Five new phenolic glycosides, cucurbitosides A-E, were isolated from the seeds of *Cucurbita moschata* by Koike *et a.*, (2005). During the screening of a variety of plant sources for their anti-obesity activity, it was found that a water-soluble extract, named PG105, prepared from stem parts of *Cucurbita moschata*, contains potent anti-obesity activities (Choi *et al.*, 2007).

Among the monosaccharides in pumpkin glucose (21.7%) and glucuronic acid (18.9%) were identified to be the main monosaccharides, followed by galactose (11.5%), arabinose (9.8%), xylose (4.4%), and rhamnose (2.8%). They were also found to have good antioxidant activity (Yang *et al.*, 2007).

Azevedo and Rodriquez (2007) reported that the principal carotenoids in *C. moschata* were beta-carotene and alpha-carotene, while it is lutein and beta-carotene that are dominate in *C. maxima* and *C. pepo*.

Rodriguez *et al.*, (1996) isolated delta 5 and delta 7 sterols having saturated and unsaturated side chain from the sterol fraction of seed oil from commercial *Cucurbita moschata* Dutch and they also elucidated its structure.

Health Benefits of Pumpkin Based on Traditional Knowledge/Folklore Practices

• It is low calorie vegetable, and is rich in dietary fiber, antioxidants, minerals, vitamins. It is recommended by dieticians in controlling cholesterol and reducing weight.

- With 7384 µg per 100 g, it is one of the vegetables in the aphaninae family with highest levels of vitamin A, providing about 246% of RDA. Vitamin A is a powerful natural antioxidant and is required by body for maintaining the integrity of skin and mucus membranes. It is also an essential vitamin for vision.
- It is recommended for oral cavity diseases, including oral cancer. It is said to prevent aging and keep skin glowing.
- Pumpkin is believed to be effective against depression, hypertension, and heart ailments.
- Being rich in fibres and pectin, it is recommended for lowering the body fat, fight high cholesterol problem and easy bowel movement.
- The presence of zinc in pumpkins boosts the immune system and also improves the bone density.
- They have been known to reduce inflammation, without causing the side effects of anti-inflammatory drugs.

Therapeutic Properties of Pumpkin

Anti-cancer: The anti-cancer properties of pumpkin are reported to be due to the presence of ribosome inactivating aphanin (RIPs). Plant RIPs (Ribosome Inactivating proteins) are the majority member of RIP superfamily, and widely distribute in different plants and various tissues, in some cases, even a large quantity in the seeds (Barbieri *et al.*, 1993; Liu *et al.*, 2002). They are shown to have diversified biological functions, such as antiviral (anti-HIV), antifungal and insecticidal properties (Au *et al.*, 2000; Nielsen *et al.*, 2001; Zhou *et al.*, 2000; Stebbing *et al.*, 2003). There are three types of RIPs based on their structure, and Moschatin from the seeds of pumkin is Type I RIP, first reported by Heng *et al.*, (2003) and demonstrated that it has a capability to selectively kill human melanoma cells. Similary, cucurmosin a type I RIP (isolated from pumpkin) is also reported to induce the apoptosis of BxPC-3 pancreatic cancer cells via the PI3K/Akt/mTOR signaling pathway (Zhang *et al.*, 2011).

Anti-hyperlipidemia: The extracts of pumpkin was found to inhibit the HMG-CoA reductase activity thereby reducing the cholesterol levels in animal models. Its effect was found to be equal to that of pravastatin, a drug used for management of hyperlipidemia (Duangajai *et al.*, 2010).

Ani-Obesity: During the screening of a variety of plant sources for their anti-obesity activity, it was found that a water-soluble extract, named PG105, prepared from stem parts of *Cucurbita moschata*, contains potent anti-obesity activities in a high fat diet-induced obesity mouse model. To understand the underlying mechanism at the molecular level, when the effects of PG105 were examined on the expression of the genes involved in lipid metabolism by Northern blot analysis, it was found that in the liver of PG105-treated mice, the mRNA level of lipogenic genes such as SREBP-1c and SCD-1 was decreased, while that of lipolytic genes such as PPARalpha, ACO-1, CPT-1, and UCP-2 was modestly increased. Therefore, PG105 from *Cucurbita* *moschata* is believed to have great potential as a novel anti-obesity agent in that both inhibition of lipid synthesis and acceleration of fatty acid breakdown (Choi *et al.*, 2007).

Anti-oxidant Activity: The various kinds of monosaccharides in pumpkin act as antioxidants and protect the cell against damage which has been demonstrated by its effective inhibition of hydrogen peroxide induced decrease of cell viability, SOD activity, lactate dehydrogenase leakage, and malondialdehyde formation, and glutathione depletion in cultured mouse peritoneal macrophages (Yang *et al.*, 2007). Nkosi *et al.*, (2006) also demonstrated that pumpkin seed protein isolate administration was effective in alleviating the detrimental effects associated with protein malnutrition and carbon tetrachloride intoxication. It is therefore, apparent that pumpkin seed protein isolate has components that have antiperoxidative properties.

Anti-fungal Activity: Cucurmoschin, a novel anti-fungal peptide found in pumpkin inhibited mycelial growth in the fungi *Botrytis aphanin, Fusarium oxysporum* and *Mycosphaerella oxysporum*. It inhibited translation in a cell-free rabbit reticulocyte lysate system (Wang and Ng, 2003).

Macular Degeneration: Fruits and vegetables rich in lutein and zeaxanthin decrease the risk for age related macular degeneration. Among the vegetable's pumpkin is rich in leutin, zeaxanthin, neozanthin and violazanthin and therefore, demonstrated reduced risk of macular degeneration (Sommerburg *et al.*, 1998).

References

Au, T.K., Collins, R.A., Lam, T.L., Ng, T.B., Fong, W.P., and Wan, D.C. 2000. The plant ribosome inactivating proteins luffin and saporin are potent inhibitors of HIV-1 integrase. FEBS Lett., (2000), 471:169-72.

Azevedo-Meleiro, C.H., and Rodriguez-Amaya, D.B. 2007. Qualitative and quantitative differences in carotenoid composition among Cucurbita moschata, Cucurbita maxima, and Cucurbita pepo. J Agric Food Chem., (2007), 55(10):4027-33.

Barbieri, L., Battelli, M.G., and Strpe, F. 1993. Ribosome-inactivating proteins. From plants. Biochim Biophys Acta., (1993), 154:237-82.

Bose, T.K. and Som, M.G. 1986. Vegetable crops in India. Naya Prokash, Calcutta.

Choi, H., Eo, H., Park, K., Jin, M., Park, E.J., Kim, S.H., Park, J.E., and Kim, S. 2007. A water-soluble extract from *Cucurbita moschata* shows anti-obesity effects by controlling lipid metabolism in a high fat diet-induced obesity mouse model. Biochem Biophys Res Commun., (2007), 359(3):419-25.

Duangjai, A., Ingkaninan, K., and Limpeanchob, N. 2010. Potential mechanisms of hypocholesterolaemic effect of Thai spices/dietary extracts. Nat Prod Res., (2011), 25(4):341-52.

Gopalan, C., Sastri, B.V.R., and Balasubramanian, S.C. 1985. Nutritive Value of Indian Foods. National Institue of Nutrition, ICMR, Hyderabad, India.

Xia, Heng Chuana., Feng Li., Zhen Li,, and Zu Chuan Zhang. 2003. Purification and characterization of Moschatin, a novel type I ribosome-inactivating protein from the mature seeds of pumpkin *(Cucurbita moschata)*, and preparationof its immunotoxin against human melanoma cells. Cell Research, (2003), 13(5):369-374.

Jiang, Z., and Du, Q. 2011.Glucose-lowering activity of novel tetrasaccharide glyceroglycolipids from the fruits of *Cucurbita moschata*. Bioorg Med Chem Lett., (2011), 21(3):1001-3.

Koike, K., Li, W., Liu, L., Hata, E., and Nikaido, T. 2005. New phenolic glycosides from the seeds of Cucurbita moschata. Chem Pharm Bull (Tokyo)., (2005), 53(2):225-8.

Li, F.S., Dou, D.Q., Xu, L., Chi, X.F., Kang, T.G., and Kuang, H.X. 2009. New phenolic glycosides from the seeds of *Cucurbita moschata*. J Asian Nat Prod Res., (2009), 11(7):639-42.

Liu, R.S., Yang, J.H., and Liu, W.Y. 2002. Isolation and enzymatic characterization of lamjapin, the first ribosome-inactivating protein fromcryptogamic algal plant (*Laminaria japonica* A). Eur J Biochem., (2002), 269(19):4746-52.

Nielsen, K., Payne, G.A., and Boston, R.S. 2001.Maize ribosome-inactivating protein 1 has antifungal activity against *Aspergillus flavus* and *Aspergillus nidulans.* Mol Plant-Microbe Interact., (2001), 14:164- 72.

Nkosi, C.Z., Opoku, A.R., and Terblanche, S.E. 2006. Antioxidative effects of pumpkin seed (*Cucurbita pepo*) protein isolate in CCl4-induced liver injury in low-protein fed rats. Phytother Res., (2006), 20(11):935-40.

Rodriguez, J.B., Gros, E.G., Bertoni, M.H., Cattaneo, P. 1996. The sterols of *Cucurbita moschata* ("calabacita") seed oil. Lipids, (1996), 31(11):1205-8.

Sommerburg, O., Jan, E.E.K., Alan, C. B., Frederik, J. .G. M., and van Kuijk. 1998. Fruits and vegetables that are sources for lutein and zeaxanthin: the macular pigment in human eyes. Br J Ophthalmol., (1998), 82:907–910.

Stebbing, J., Patterson, S., and Gotch F. 2003. New insights into the immunology and evolution of HIV. Cell Res., (2003), 13(1):1-7.

Wang, H.X., and Ng, T.B. 2003. Isolation of cucurmoschin, a novel antifungal peptide abundant in arginine, glutamate and glycine residues from black pumpkin seeds., (2003), 24(7):969-72.

Yadav, M., Jain, S., Tomar, R., Prasad, G.B., and Yadav, H. 2010. Medicinal and biological potential of pumpkin: An updated review. Nutr Res Rev., (2010), 23(2):184-90.

Yang, X., Zhao. Y., and Lv, Y. 2007. Chemical composition and antioxidant activity of an acidic polysaccharide extracted from *Cucurbita moschata* Duchesne ex Poiret. J Agric Food Chem., (2007), 55(12):4684-90.

Zhang, B., Huang, H., Xie, J., Xu, C., Chen, M., Wang, C., Yang, A., and Yin, Q. 2011. Cucurmosin induces apoptosis of BxPC-3 human pancreatic cancer cells via inactivation of the EGFR signaling pathway. Oncol Rep., (2011) 27(3):891-897. Doi: 10.3892/or.2011.1573.

Zhou, X., Li, X.D., Yuan, J.Z., Tang, Z.H., and Liu, W.Y. 2000. Toxicity of cinnamomin a new type II ribosome-inactivating protein to bollworm and mosquito. Insect Biochem Mol Biol., (2000), 30(3): 259-64.

2.10. Musk Melon (*Cucumis melo* L.)

Musk melon belongs to the family of Cucurbitaceae. The species *Cucumis melo* is a large taxon, encompassing many horticultural groups/varieties. Among these *Cucumis melo* L., Reticulata group includes, muskmelon, Persian melon or netted melon; while Cantaloupensis group includes cantaloupe and Inodorous group includes honeydew, winter or casaba melon. The muskmelon is native to Near East (probably Persia or Iran). All but a handful of culinary melon varieties belong to the species *Cucumis melo* L. Muskmelon (*Cucumis melo*) is said to be native to tropical Africa, most specifically to eastern region and south of Sahara Desert. Melons are used as a dessert or breakfast fruit or in fruit salads, punches, jellies, *etc*. Nutritionally muskmelon is low in calories and a good source of vitamin A and vitamin C.

Indians have cultivated several varieties that adapt to a wide range of climates. Approximately 10 commercial varieties of musk melon have importance in India. Musk melons season is during the summer months from April to July. A few off-season melons may make their way onto the markets throughout the year. Muskmelon is grown in states of Uttar Pradesh, Punjab, Andhra Pradesh, Rajasthan, Madhya Pradesh, Bihar and Karnataka.

Cultivars/Varieties: Arka Jeet, Arka Rajhans, Pusa Sharbati, Pusa Madhuras, Hara Madhu, Durgapur Madhu, Punjab Sunehri and Punjab Hybrid are some of the popular cultivars grown in India (Bose and Som, 1986). Other noted locally grown desert cultivars are Lucknow Safeda, Baghpat melon, Jaunpuri Netted, Mau melon of Azamgarh, Tonk melon of Rajasthan, and Sharbat-e-Anar of Andhra Pradesh.

The third but less widely cultivated species of this genus is *Cucumis anguira*, familiarly called as West Indian Gerkhin. At present there are about 20 units in the country producing and exporting gherkins in India. Gherkins, is one of the crops in India which are cultivated exclusively for exports as their consumption within the country is almost nil.

Nutritional Composition of Musk melon

Constituents	Quantity (per 100g)	Constituents	Quantity (per 100g)
Moisture	95.20 g	Phosphorus	14 mg
Carbohydrates	3.50 g	Iron	1.4 mg
Protein	0.30 g	Potassium	487 mg
Fat	0.20 g	Thiamine	0.11 mg
Minerals	0.40 g	Riboflavin	0.08 mg
Fibre	0.40 g	Niacin	0.30 mg
Energy	17 Kcal	Vitamin C	26 mg
Calcium	32 mg	Carotene	169 µg

(Gopalan *et al.*, 1985)

Phytochemicals

The major phytochemicals found in muskmelon are vitamin C, beta-carotene, minerals like potassium, phosphorus and iron. Besides, there are several secondary metabolites and compounds that have health benefits. These are being extensively studied by various groups in different parts of the world. Six fragrant ingredients were identified in fully-ripened Katsura-uri (Japanese pickling melon; *Cucumis melo* var. conomon). Four of them were sulfur-containing compounds [methylthioacetic acid ethyl ester (MTAE), acetic acid 2-methylthio ethyl ester (AMTE), 3-methylthiopropionic acid ethyl ester (MTPE), and acetic acid 3-methylthio propyl ester (AMTP)]; and the others were benzyl acetate and eugenol. These chemicals exhibited anticarcinogenic properties (Nakamura *et al.*, 2010; Nakamura *et al.*, 2008)). Enzymatic antioxidants like superoxidismutase (SOD) were found to be present in various amounts in different varieties of melons (Lester *et al.*, 2009). Non-enzymatic antioxidants like beta-carotene which act as free radical scavenging agents were also found abundantly in some types. The phytochemical analysis indicated presence of a high number of polyphenols and ascorbic acid too in muskmelon peel extracts suggesting that the beneficial effects could be the result of the rich content of polyphenols and ascorbic acid in the peels (Parmar and Kar, 2008). Oxykine is the compound found in cantaloupe melon extract rich in vegetal superoxide dismutase (Naito *et al.*, 2005).

Studies on *C. melo* L. led to isolation of hydrocarbons (Velcheva and Donchev, 1997), peptides (Ribeiro *et al.*, 2007), fatty acids, volatile sesquiterpenes (Lewinsohn *et al.*, 2008; Portnoy *et al.*, 2008), phenylethyl chromone derivatives, triterpenes, sterols, and one triacylglyceride (Ibrahim, 2010, 2014).

Ibrahim and Mohamed (2015) isolated a new phenylethyl chromenone, cucumin S along with five known compounds: two chromones, luteolin, quercetin, and 7-glucosyloxy-5-hydroxy-2-[2-(4-hydroxyphenyl) ethyl] from the EtOAc fraction of *Cucumis melo* var. *reticulatus* seeds. Their structures were determined and the isolated compounds 1–6 were assessed for their antioxidant activity using DPPH assay. Three of these isolated compounds showed potent activities compared to propyl gallate at concentration 100 M.

Health Benefits of Musk melon & Cantaloupes based on Traditional Knowledge/Folklore practices

- Muskmelon comprises of a significant amount of dietary fiber, making it good for those suffering from constipation.
- Since the fruit is low in sugar or calories, it serves as a good snack for dieting.
- The potassium present in muskmelons makes them quite helpful in the lowering of blood pressure and regulating heartbeat and, possibly, preventing strokes.
- Being rich in Vitamin C, an antioxidant, the fruit is believed to be good for preventing heart diseases, cancer and other chronic ailments.
- Muskmelons have been found to have sedative properties, making them beneficial for the people who are suffering from insomnia.
- Regular consumption of muskmelon juice can help treat lack of appetite, acidity, ulcer and urinary tract infections.
- Since it has high water content, the fruit can help reduce the heat in the body and thus, prevent heat-related disorders.

- The fruit is a good source of vitamin A and thus, helps maintain healthy skin.
- The folic acid present in muskmelons helps create healthy foetuses (in pregnant women) and can even prevent cervical cancer and osteoporosis. It also serves as a mild antidepressant.
- Being good source of beta-carotene, it is helpful in reducing the macular denegeration.
- Muskmelon & Cantaloupe are excellent source of vitamin A, vitamin C and vitamin E which is potent antioxidant and is helpful in boosting the immune system.
- Potassium present in muskmelon works closely with sodium to help keep blood pressure stable and prevent muscles from cramping. In addition, potassium helps maintain a proper electrolyte balance in extracellular fluid, and in doing so reduces water retention.
- It is also a very good source of dietary fiber, vitamin B1, vitamin B3 (niacin), vitamin B6, folate, magnesium, and vitamin K.

Therapeutic Properties of Musk melon

Researchers evaluated the effect of consuming the antioxidant vitamins A, C, and E and carotenoids through fruits and vegetables on the development of early ARMD or neovascular ARMD, a more severe form of the illness associated with vision loss. Fruit intake was definitely protective against the severe form of vision-destroying disease. The newly identified muskmelon frangrants like MTAE and AMTP possessed antimutagenic activity as determined by their ability to inhibit the UV-induced mutation in repair-proficient E. coli B/r WP2 (Nakamura *et al.*, 2010).

Vouldoukis *et al.*, (2004) evaluated *in vitro* and *in vivo* the antioxidant and anti-inflammatory properties of a cantaloupe melon (*Cucumis melo* LC., Cucurbitaceae) and postulated that Cucumil Melo extract (CME) inhibited the production of peroxynitrite, strengthening the antioxidant properties of the CME rich in SOD activity. The production of the pro- and anti-inflammatory cytokines, namely TNF-alpha and IL-10, being conditioned by the redox status of macrophages, they also evaluated the effect of CME and HI-CME on the IgG1IC-induced cytokine production. Their data demonstrated that, in addition to its antioxidant properties, the anti-inflammatory properties of the CME extract were principally related to its capacity to induce the production of IL-10 by peritoneal macrophages.

Rats, treated with the *Cucumis melo* peel extracts reversed the CCT-diet induced increase in the levels of tissue LPO, serum lipids, glucose, creatinine kinase-MB and decrease in the levels of thyroid hormones and insulin indicating their potential to ameliorate the diet induced alterations in serum lipids, thyroid dysfunctions and hyperglycemia/diabetes mellitus. This indicated their usefulness as food supplements in treating some medical conditions (Parmar and Kar, 2008).

References

Bose, T.K. and Som, M.G. 1986. Vegetable crops in India. Naya Prokash, Calcutta.

Gopalan, C., Sastri, B.V.R. and Balasubramanian, S.C. 1985. Nutritive Value of Indian Foods. National Institue of Nutrition, ICMR, Hyderabad, India.

Ibrahim, S.R.M. 2010. New 2-(2-phenylethyl) chromone derivatives from the seeds of *Cucumis melo* L var. *reticulatus*. Nat. Prod. Commun., 5:403–406.

Ibrahim, S.R.M. 2014. New chromone and triglyceride from *Cucumis melo* seeds. Nat. Prod. Commun., 9: 205–208.

Ibrahim, S.R.M., and Mohamed, G.A. 2015. Cucumin S, a new phenylethyl chromone from *Cucumis Melo* var. *reiculatus* seeds. Revistra Brasileira de Farmacognosia, (2015), 25(5): 462-464.

Lester, G.E., Jifon, J.L., Crosby, K.M. 2009. Superoxide dismutase activity in mesocarp tissue from divergent Cucumis melo L. genotypes. Plant Foods Hum Nutr., (2009), 64(3):205-11.

Lewinsohn, E., Portnoy, V., Benyamini, Y., Bar, E., Harel-Beza, R., Gepstein, S., Giovannoni, J.J., Schaffer, A.A., Burger, Y., Tadmor, Y., and Katzir, N. 2008. Sesquiterpene biosynthesis in melon (*Cucumis melo* L.) rinds. In: Pitrat, M. (Ed.), Cucurbitaceae Proceedings of the Ixth EUCARPIA Meeting on Genetics and Breeding of Cucurbitaceae. INRA, Avignon, France, pp. 249–255.

Naito, Y., Akagiri. S., Uchiyama, K., Kokura, S., Yoshida, N., Hasegawa, G., Nakamura, N., Ichikawa, H., Toyokuni,S., Ijichi, T., Yoshikawa, T. 2005. Reduction of diabetes-induced renal oxidative stress by a cantaloupe melon extract/gliadin biopolymer, oxykine, in mice. Biofactors, (2005), 23(2):85-95.

Nakamura, Y., Watanabe, S., Kageyama, M., Shirota, K., Shirota, K., Amano, H., Kashimoto, T., Matsuo, T., Okamoto, S., Park, E.Y., Sato, K. 2010. Antimutagenic; differentiation-inducing; and antioxidative effects of fragrant ingredients in Katsura-uri (Japanese pickling melon; *Cucumis melo* var. conomon). Mutat Res., (2010), 703(2):163-8.

Nakamura, Y., Nakayama, Y., Ando, H., Tanaka, A., Matsuo, T., Okamoto, S., Upham, B.L., Chang, C.C., Trosko, J.E., Park, E.Y., Sato, K. 2008. 3-Methylthiopropionic acid ethyl ester, isolated from Katsura-uri (Japanese pickling melon, *Cucumis melo* var. conomon), enhanced differentiation in human colon cancer cells. J Agric Food Chem., (2008), 56(9):2977-84.

Parmar, H.S., and Kar, A. 2008. Possible amelioration of atherogenic diet induced dyslipidemia, hypothyroidism and hyperglycemia by the peel extracts of *Mangifera indica, Cucumis melo* and *Citrullus vulgaris* fruits in rats, (2008);33(1):13-24.

Portnoy, V., Benyamini, Y., Bar, E., Harel-Beza, R., Gepstein, S., Giovannoni, J.J., Schaffer, A.A., Burger, Y., Tadmor, Y., Lewinsohn, E., and Katzir, N. 2008. The molecular and biochemical basis for varietal variation in sesquiterpene content in melon (*Cucumis melo* L.) rinds. Plant Mol. Biol., 66: 647–661.

Ribeiro, S.F.F., Agizzio, A.P., Machado, O.L.T., Neves-Ferreira, A.G.C., Oliveira, M.A., Fernandes, K.V.S., Carvalho, A.O., Perales, J., Gomes, V.M.A., 2007. A new peptide of melon seeds which shows sequence homology with vicilin: partial characterization and antifungal activity. Sci. Hortic., 111: 399–405.

Velcheva, M.P., and Donchev, C. 1997. Isoprenoid hydrocarbons from the fruit of extant plants. Phytochemistry, 45: 637–639.

Vouldoukis, I., Lacan, D., Kamate, C., Coste, P., Calenda, A., Mazier, D., Conti, M., and Dugas, B. 2004. Antioxidant and anti-inflammatory properties of a *Cucumis melo* LC. Extract rich in superoxide dismutase activity. J Ethnopharmacol., (2004) 94(1):67-75.

2.11. Watermelon (*Citrullus lanatus* L.)

Watermelon belongs to Citrullus genus in Cucurbitaceae family. *Citrullus vulgaris* or cultivated watermelons is divided into vars. Lanatus and Citroides, the latter comprising the citron or preserving melon. Watermelon is a vine-like (scrambler and trailer) flowering plant originally from southern Africa.

Watermelon is used fresh as a dessert fruit and alone or together with other fruits in fruit salads. The watermelon rind can be pickled as a preserve. Watermelon is low in calories and is a good source of vitamin C and vitamin A. As a member of the Cucurbitaceae family, the watermelon is related to the cantaloupe, squash and pumpkin, other plants that also grow on vines on the ground. Watermelons can be round, oblong or spherical in shape and feature thick green rinds that are often spotted or striped. They range in size from a few pounds to upward of ninety pounds. Although watermelons can now be found in the markets throughout the year, the season for watermelon is in the summer when they are sweet and of the best quality. Watermelon is grown in states West Bengal, Orissa, Andhra Pradesh, Uttar Pradesh and Karnataka

Cultivars/Varieties: Ashahi Yamata, Sugar Baby, Arka Jyoti, Arka Manik, Durgapur Meetha, Durgapur Kesar are some varieties that had been under cultivation for very long period of time. Several new varieties have been released recently like Kiran, Vishala, Saraswathi, Mithila and Priya are some varieties released by Known-You Seed India Pvt Ltd. Arka Muthu is the latest variety released from IIHR, Bangalore.

Nutritive Components of Watermelon

Constituents	Quantity (per 100g)	Constituents	Quantity (per 100g)
Moisture	91.5 g	Phosphorus	11 mg
Carbohydrates	7.75 g	Iron	0.24 mg
Protein	0.61 g	Potassium	112 mg
Fat	0.15 g	Thiamine	0.033 mg
Minerals	0.30 g	Riboflavin	0.021 mg
Fibre	0.40 g	Niacin	0.178 mg
Energy (Kcal)	16 Kcal	Vitamin C	8.1 mg

(Goplan *et al.*, 1985; http://healthmad.com/nutrition/10-health-benefits-of-watermelon)

Phytochemicals in Watermelon

The watermelon (*Citrullus lanatus*) seeds are highly nutritive and contain large amount of proteins and many beneficial minerals such as magnesium, calcium, potassium, iron, phosphorous, zinc *etc.*, (Yadav *et al.*, 2011). A low molecular weight vicilin-like glycoprotein was purified using chromatographic methods followed by SDS-PAGE and MALDI-TOF/MS identification. Being a member of aphaninae family, watermelon also contains cucurbitacin E. Wani *et al.*, (2010) showed that globulin was the major protein (≥500 g kg (-1)) present in watermelon seed meal, followed by albumin and glutelin. These proteins had good functional properties based on which they suggested its potential use in food formulations.

Mostly concentrated in the rind and white fleshy parts, Citrulline is a non-essential amino acid that supports the body stress, muscle fatigue. Citrulline is used in the nitric oxide system and has potential antioxidant and vasodilatation roles. Watermelon has the highest level of lycopene found in fruits and vegetables. It has good amount of Vitamin B6 which support the immune system and increase metabolism. Vitamin B6 has been specifically noted in studies to help ward off anxiety, depression and high blood pressure. It helps brain function and helps the body convert protein to energy. It also has Vitamin C which boosts immunity. Watermelons also have Vitamin A and carotenoids which kepps the body free from toxins. It is also reported to have lutein which slow age-related macular degeneration in eyes. The essential minerals found abundantly in watermelons are calcium, magnesium, phosphorus and potassium. They also contain smaller amounts of iron and zinc. They also have good amount of fibers which not only lowers cholesterol but keeps the digestive tract running smoothly and by giving a full feel of the stomach its consumption prevents overeating of other calorie dense foods like carbohydrates and fats.

Health Benefits of Watermelon based on Traditional Knowledge/Folk-lore practices

- Watermelon helps to reduce the inflammation that contributes to conditions like asthma, atherosclerosis, diabetes, colon cancer, and arthritis.
- Watermelon is also a very concentrated source of the carotenoid, lycopene. Lycopene has been extensively studied for its antioxidant and cancer-preventing properties.
- Watermelon is good for eyes as the carotenoids present in it helps in controlling macular degeneration.
- It is high in citrulline, an amino acid our bodies use to make another amino acid, arginine, which is used in the urea cycle to remove ammonia from the body, and by the cells lining our blood vessels to make nitric oxide. Arginine has been shown to improve insulin sensitivity in obese type 2 diabetic patients with insulin resistance.
- **Asthma:** The antioxidants, such as Citrulline in watermelon can reduce toxins in the body. Some of these toxins are believed to trigger asthma attacks (Edwards *et al.*, 2003; Collins *et al.*, 2007).
- **Arthritis:** Watermelon is rich in Vitamin C and Beta-Carotenes. Both have anti-inflammatory properties that can help relieve the symptoms of arthritis.
- **Bladder Problems:** Containing 92% water, this melon acts as a natural diuretic that cleanses your kidney and bladder.
- **High Cholesterol:** The cleansing properties of watermelon may also reduce the risk of clogged arteries. The fruit is also believed to increase HDL (good cholesterol levels).
- **Constipation:** A glass of watermelon juice can naturally eliminate waste and promote proper digestion.
- **Bloating:** Again, the diuretic effect of watermelon can eliminate water retention. It is especially recommended for women who retain fluids during menstruation or pregnancy.

- **Heart Disease:** Watermelon contains Folic acid and Citrulline that works in tandem with other essential vitamins. Some research indicates that these vitamins may reduce the risk of a heart attack, stroke and colon cancer.
- **Itchy Skin:** As the water from the melon cleanses your system, it can flush out the toxic waste that cause some forms of itchiness.
- **Prostate Cancer:** The Lycopene in watermelon has been extensively reported to reduce the risk of prostate cancer (Erhardt *et al.*, 2003).
- **Blemishes:** Watermelon can also be used as a beauty aid to reduce skin blemishes. Simply rub your skin with a small piece of watermelon. Leave the juice on for 10 minutes before rinsing.

Therapeutic Properties of Watermelon

In various parts of the world, *C. lanatus* seed extracts are used to cure cancer, cardiovascular diseases, hypertension, and blood pressure. *C. lanatus* seed extracts are also used as home remedy for edema and urinary tract problems (Yadav *et al.*, 2011). Cucurbitacin E (CE) is potentially useful in treating inflammation through the inhibition of COX and RNS but not ROS. This was demonstrated by Abdelwahab *et al.*, (2011). CE inhibited both COX enzymes with more selectivity toward COX-2. Intraperitoneal injection of CE significantly suppressed carrageenan-induced rat's paw edema.

Altas *et al.*, (2011) demonstrated that watermelon juice protects the liver, kidney and brain tissues from experimental CCl(4) toxicity in rats and that the protective effect of watermelon juice may be due to its antioxidant activity and inhibition of lipid peroxide formation. CLMT2 gene present in watermelon had an extraordinarily high activity for detoxifying hydroxyl radicals suggesting it to be responsible for its high drought tolerance. This also indicates its antioxidant potential (Akashi *et al.*, 2004).

The study conducted by Veazie *et al.*, (2008) showed that subjects drinking six cups of watermelon juice per day had increased levels of arginine and lycopene content in their plasma. A diet enriched with citrulline/arginine or watermelon reduced glucose levels and improved aortic flexibility in an animal model study. They suggested that high amounts of lycopene in red fleshed watermelon may be useful in blocking free radical damage, while the citrulline may improve vascular health.

References

Abdelwahab, S.I., Hassan, L.E., Sirat, H.M., Yagi, S.M., Koko, W.S., Mohan, S., Taha, M.M., Ahmad, S., Chuen, C.S., Narrima, P., Rais, M.M., and Hadi, A.H. 2011. Anti-inflammatory activities of cucurbitacin E isolated from *Citrullus lanatus* var. citroides: role of reactive nitrogen species and cyclooxygenase enzyme inhibition. Fitoterapia, (2011), 82(8) :1190-7.

Akashi, K, Nishimura, N., Ishida. Y., and Yokota, A. 2004. Potent hydroxyl radical-scavenging activity of drought-induced type-2 metallothionein in wild watermelon. Biochem Biophys Res Commun., (2004), 323(1):72-8.

Altaş, S., Kızıl, G., Kızıl, M., Ketani, A., and Haris, P.I. 2011. Protective effect of Diyarbakır watermelon juice on carbon tetrachloride-induced toxicity in rats. Food Chem Toxicol., (2011), 49(9):2433-8.

Collins, J, K., Wu, G., Perkins-Veazie, P., Spears, K., Claypool, P.L., Baker, R.A., and Clevidence, B.A. 2007. Watermelon consumption increases plasma arginine concentrations in adults. Nutrition, (2007), 23(3):261-6. 2007.

Edwards, A.J., Vinyard, B.T., Wiley, E.R., et al., 2003. Consumption of watermelon juice increases plasma concentrations of lycopene and beta-carotene in humans. J Nutr., (2003), 133(4):1043-50 2003.

Erhardt, J.G., Meisner, C., Bode, J.C., Bode, C. 2003. Lycopene, beta-carotene, and colorectal adenomas. Am J Clin Nutr., (2003), 78(6):1219-24. 2003.

Gopalan, C., Sastri, B.V.R. and Balasubramanian, S. C. 1985. Nutritive Value of Indian Foods. National Institue of Nutrition, ICMR, Hyderabad, India.

Perkins, V.P.M., Collins, J.K., Wu, G., Clevidence, B.A. 2008. Watermelon, phytochemicals, and health. Proceedings of the United States-Japan Cooperative Program in Natural Resources. 37th Annual Meeting, August 24-28, 2008, Chicago, Illinois. P. 138-139.

Wani, A.A., Sogi, D.S., Singh, P., Wani, I.A., Shivhare, U.S. 2010. Characterisation and functional properties of watermelon (Citrullus lanatus) seed proteins. J Sci Food Agric., (2011), 91(1):113-21. Doi: 10.1002/jsfa.4160. Epub 2010 Sep 7.

Yadav, S., Tomar, A.K., Jithesh, O., Khan, M.A., Yadav, R.N., Srinivasan, A., Singh, T.P., and Yadav, S. 2011. Purification and partial characterization of low molecular weight vicilin-like glycoprotein from the seeds of Citrullus lanatus. Protein J., (2011), 30(8):575-80.

3. Cole Crops

Cole crops are a group of vegetables belonging to the family Brassicaceae (Syn: Cruciferae) and include cabbage, cauliflower, knoll khol, kale, Brussel sprouts, sprouting broccoli, *etc*. Cole crops are found spread all over Europe from the Mediterranean region, which is supposed to be the centre of origin. In the ancient Egyptian and Asian civilization, there was mention of cole crops and also by Aristotle in his writings on 'Enquiry into plants' there was mention about three types of coles. It is believed that it was first collected for food at the beginning of Neolithic times before being cultivated, and the cultivated forms of cabbage group like, cabbage, savoy, kales, collard, broccoli, Brussels sprouts, cauliflower and kohlrabi gradually originated from wild species through mutation, human selection and adaptation.

Among the cole crops in India cauliflower and cabbage are the two most important crops while Kohlrabi is grown very less. In India it is cultivated in states of Uttar Pradesh, Karnataka, Maharashtra, Bihar, West Bengal, Punjab and Haryana. It is also grown in milder climates in southern India like Ooty, Bangalore, *etc*.

3.1. Cabbage (*Brassica oleracea* L.f. *alba*)

Cabbage, *Brassica oleracea Linne* (Capitata Group) is a popular vegetable belonging to the family Brassicaceae (or Cruciferae) and is used as a leafy green vegetable. It is in the same genus as the turnip – *Brassica rapa*. Cabbage leaves often have a delicate, powdery, waxy coating called *bloom*. The occasionally sharp or bitter taste of cabbage is due to glucosinolate(s). Cabbages are also a good source of riboflavin.

In India cabbage is widely grown in the states of Uttar Pradesh, Orissa, Bihar, Assam, West Bengal, Maharashtra and Karnataka.

Cultivars/Varieties: Cabbage cultivars are classified on the basis of colour of head and maturity. In India white cabbage is important and these are available in pointed, round and flat shapes. Of these round cabbages are preferred. Common cabbage varieties grown are Golden Acre, Pride of India, Copenhagen Market, Red Acre, Pusa Drum Head, Pride of India, Pusa Agethi, Pusa Muktha, KGMR-1, Kateri, Hardam and Nakshatra.

Uses

The eaten part of the cabbage is the leafy head comprising mainly the immature leaves. Cabbage is used in a variety of dishes for its naturally spicy flavor. It is widely consumed raw, cooked, or preserved in a great variety of dishes. It is one of the ingredients in Chinese fast foods, burgers and sandwiches. The pickled cabbage called 'sauerkraut' is one of the oldest foods known in European countries. Chinese 'suan cai' and Korean 'kimchi' are produced using the related Chinese cabbage.

Nutritional Components of Cabbage

Constituents	Quantity (per 100g)	Constituents	Quantity (per 100g)
Carbohydrates	5.80 g	Zinc	0.18 mg
Proteins	1.28 g	Energy	25 Kcal
Fats	0.10 g	Thaimine	0.061 mg
Dietary Fibre	2.50 g	Riboflavin	0.040 mg
Minerals	0.6 g	Niacin	0.234 mg
Calcium	40 mg	Pantothenic acid	0.212 mg
Phosphorus	26 mg	VitaminB6	0.124 mg
Iron	0.47 mg	Folic acid	53 µg
Magnesium	12 mg	Vitamin C	36.6 g
Potassium	170 mg		

(Gopalan et al., 1985)

Cabbage is an excellent source of vitamin C. It also contains significant amounts of glutamine, an amino acid that has anti-inflammatory properties. Cabbage can also be included in dieting programs, as it is a low calorie food.

Phytochemicals

Being a member of Cruciferaceae family, cabbage is rich in sulphur compounds. It is good source of dietary fibre and proteins besides considerable amounts of calcium, phosphorus and potassium. Singh et al., (2006 & 2007) compared eighteen different cabbage cultivars to see the variation in their antioxidant phytonutrients. The principal nutrients identified were Vitamin C, β-carotene, lutein, vitamin E (dl-α-tocopherol) and phenols.

Brassiparin represents one of the few antifungal proteins reported to date from *Brassica* species. Its antifungal activity has pronounced pH stability and thermostability. Brassiparin exhibits other exploitable activities such as antiproliferative activity toward liver and breast cancer cells and inhibitory activity toward HIV-reverse transcriptase (Lin and Ng, 2009). Cabbage is a source of indole-3-carbinol, a chemical which boosts DNA repair in cells and appears to block the growth of cancer cells. Indole-3-carbinol (I3C), induces a G1 cell-cycle arrest of human breast cancer cells, although the direct cellular targets that mediate this process are unknown (Nguyen et al., 2008). The compound is also used as an adjuvant therapy for recurrent respiratory papillomatosis, a disease of the head and neck caused by human papillomavirus (usually types 6 and 11) that causes growths in the airway that can lead to death. Boiling reduces anti-cancer properties.

Health Benefits of Cabbage based on Traditional Knowledge/Folklore practices

- In European folk medicine, cabbage leaves are used to treat acute inflammation. A paste of raw cabbage is placed in a cabbage leaf and wrapped around the affected area to reduce discomfort. Some claim it is effective in relieving painfully engorged breasts in lactating women.

- Fresh cabbage juice has been shown to promote rapid healing of peptic ulcers. Cabbage is also known for slowing down growing cancer cells.
- The phytochemicals in cabbage protect the body from free radicals that can damage the cell membranes.
- Cabbage is considered good for body detoxification.
- Cabbage may lower the incidence of cancer, especially in the lung, stomach and colon prostate. The juice of fresh raw cabbage has been proven to heal stomach ulcer.
- Cabbage is a muscle builder, blood cleanser and also good to improve eye vision.
- Cabbage is rich in iron and sulfur and is effective in treating fungal and bacterial infections.
- Cabbage can lower serum cholesterol. Red Cabbage has more phytonutrients than the green cabbage. The vitamin C content of red cabbage is 6-8 times higher than that of the green cabbage.

Therapeutic properties of Cabbage

The principal health benefits of cabbage are derived from the antioxidant properties of phytochemicals like Vitamin C, β-carotene, Lutein, Vitamin E and phenols. Cabbage contain 'Sulforaphane', a substance that can increase the production of antioxidant and detoxification enzymes. Sulforaphane works by stimulating the production of glutathione, the body's most important internally produced antioxidant which plays a role in liver detoxification. Red cabbage contain anthocyanin (red pigment/color) is an antioxidant that can help protect brain cells, thus can help prevent Alzheimer's disease.

Consumption of *Brassica* (cruciferous) vegetables, such as cabbage, is directly associated with decreased risk of reproductive tissue cancers in humans (Lopez and Diamandis, 1998; Higdon *et al.*, 2007) and reduced tumor incidence in experimental animals (Grubbs, 1995). The studies implicate the existence of specific biologically active phytochemical, indole-3-carbinol (I3C), that is a potent chemotherapeutic agent. I3C is a promising molecule naturally derived from glycobrassicin in *Brassica* vegetables, which has been shown to exhibit potent anticarcinogenic properties in a wide range of cancers such as lung, liver, colon, cervical, endometrial, prostate, and breast cancer (Aggarwal and Ichikawa, 2005; Weng *et al.*, 2008; Kim and Milner, 2005).

Epidemiological studies give evidence that cruciferous vegetables (CF) protect humans against cancer, and also results from animal experiments show that they reduce chemically induced tumor formation. These properties have been attributed to alterations in the metabolism of carcinogens by breakdown products of glucosinolates, which are constituents of cucifers. These results were further confirmed from the findings of Steinkellner *et al.*, (2001).

Cancer Prevention: One of the American Cancer Society's key dietary recommendations to reduce the risk of cancer is to include cruciferous vegetables such as cabbage. Studies have indicated that increased cabbage intake may inhibit the metastatic capacity of breast cancer. Other research data provides strong evidence for a substantial protective effect of cruciferous vegetable consumption on lung cancer. Cruciferous vegetable intake has also been associated with a decrease in gastrointestinal, prostrate, and bladder cancers. The anti-cancer properties of cabbage are due to its phytochemical compounds called glucosinolates, which work primarily by increasing antioxidant defense mechanisms.

Anti-inflammatory agent: Cabbage is good source of the amino acid glutamine, which increases the body's ability to secrete human growth hormone (HGH). Glutamine also has anti-inflammatory properties and assists with immune system regulation and intestinal health.

Peptic Ulcers: Research at the Stanford University School of Medicine demonstrated that fresh cabbage juice is extremely effective in the treatment of peptic ulcers. The anti- ulcer properties of glutamine are due to the high glutamine content of cabbage.

Breast Engorgment: Breast engorgement is a painful and unpleasant condition affecting large numbers of women in the early postpartum period. Breast engorgement may inhibit the breastfeeding, leading to early breastfeeding cessation, and is associated with more serious illness, including breast infection. Studies conducted by Mangesi and Dowswell (2010) and Arora et al., (2008) have shown positive correlation between the consumption of cruciferous vegetables like cabbage and reduction in breast engorgement.

Cabbage may also act as a goitrogen. It blocks organification in thyroid cells, thus inhibiting the production of the thyroid hormones (thyroxine and tri-iodothyronine). The result is an increased secretion of thyroid-stimulating hormone (TSH) due to low thyroid hormone levels. This increase in TSH results in an enlargement of the thyroid gland (goiter) (Spence, 1933).

References

Aggarwal, B.B., and Ichikawa H. 2005. Molecular targets and anti-cancer potential of indole-3-carbinol and its derivatives. Cell Cycle, 4:1201–1215.

Arora, S., Vatsa, M., and Dadhwal, V. 2008. A Comparison of cabbage leaves vs. hot and cold compresses in the treatment of breast engorgement. Indian J Community Med., (2008), 33(3): 160–162.

Bose, T.K. and Som, M.G. 1986. Vegetable crops in India. Naya Prokash, Calcutta.

Gopalan, C., Sastri, B.V.R. and Balasubramanian, S.C. 1985. Nutritive Value of Indian Foods. National Institue of Nutrition, ICMR, Hyderabad, India.

Grubbs, C.J., Steele, V.E., Casebolt, T., Juliana, M.M., Ego, I., Whitaker, L.M., Dragnev., K.H., Kelloff, G.J., and Lubet, R.L. 1995. Chemoprevention of chemically induced mammary carcinogenesis by indole-3-carbinol. Anticancer Res., 15:709–716.

Steinkellner, H., Rabot, S., Freywald, C., Nobis, E., Scharf, G., Chabicovsky, M., Knasmüller, S., and Kassie, F. 2001. Effects of cruciferous vegetables and their constituents on drug metabolizing enzymes involved in the bioactivation of DNA-reactive dietary carcinogens.

Mutation Research/Molecular Mechanisms of Anticarcinogenesis and Antimutagenesis. Fundamental and Molecular Mechanisms of Mutagenesis, 480-481: 285-297.

Higdon, J.V., Delage, B., Williams, D.E., Dashwood, R.H. 2007. Cruciferous vegetables and human cancer risk: Epidemiologic evidence and mechanistic basis. Pharmacol Res., 55 :224–236.

Kim, Y.S., and Milner, J.A. 2005. Targets for indole-3-carbinol in cancer prevention. J Nutr Biochem., 16:65–73.

Lin, P., and Ng, T.B. 2009. Brassiparin, an antifungal peptide from Brassica parachinensis seeds. J Appl Microbiol., 106(2):554-63.

Lopez-Otin, C. and Diamandis, E.P. 1998. Breast and prostate cancer: An analysis of common epidemiological, genetic, and biochemical features. Endocr Rev., 19:365–396.

Mangesi, L., and Dowswell, T. 2010. Treatments for breast engorgement during lactation. Cochrane Database Syst Rev., 8(9):CD006946.

Nguyen, H.H., Ida, A., Gloria, A.B., David, H. H.N., Leonard F. B., and Gary, L. F. 2008. The dietary phytochemical indole-3-carbinol is a natural elastase enzymatic inhibitor that disrupts cyclin E protein processing. PNAS, 105(50): 1750-19755.

Singh, J., Upadhyay, A.K., Bahadur, A., Singh, B., Singh, P., and Rai, M. 2006. Antioxidant phytochemicals in cabbage (Brassica oleracea L. var. aphanin) Scientia Horticulturae, 108(3): 233-237.

Singh, J., Upadhyay, A.K., Prasad, K., Bahadur, A., and Rai, M. 2007. Variability of carotenes, vitamin C, E and phenolics in Brassica vegetables. Journal of Food Composition and Analysis, 20(2): 106-112.

Spence, A.W. 1933. Cabbage goiter. In Correspondence, Bri M J., (Oct 1933), page.797.

Weng, J.R., Tsai, C.H., Kulp, S.K., and Chen, C.S. 2008. Indole-3-carbinol as a chemopreventive and anti-cancer agent. Cancer Lett., 262:153–163.

3.2. Cauliflower (*Brassica oleracea var. botrytis* L.)

Cauliflower is a member of Brassicaceae family and is closely related to cabbage, broccoli, kale and turnips. Cyprus and area around Mediterranean coast is believed to be its place of origin. In spite of its late introduction, India became the leading producer of cauliflower in the world. Bihar, Uttar Pradesh, Orissa, West Bengal, Assam, Haryana and Maharashtra are the major producing states.

Cultivars/Varieties: The leading varieties under cultivation are Early Kunwari, Pant Gobhi 3, Pusa Early Synthetic (September maturing); Pusa Deepali, Pusa Katki (October maturating); Hisar 1, IIHR 101, IIHR 105, Improved Japanese, Panth Gobhi 2, PG 26, PG 35, Pusa Hybrid 2, Pusa Sharad (November maturing) and Pusa Himjyoti, Pusa Shubhra, Pusa Synthetic, Snowball-1, Pusa Snowball K-1 Snowball-16 (December maturing).

Uses

Cabbage is normally cooked and served as a vegetable. Often it is deep fried in oil and used as a Chinese dish called 'Manchurian'. Cauliflower is low in calories and is a good source of vitamin C. Similar to cabbage, cauliflower is also rich in Brassiparin. Naturally occurring chemicals (indoles, isothiocyanates, glucosinolates, dithiolethiones, and phenols) in cauliflower, and other cruciferous vegetables appear to reduce the risk of some cancers.

Nutirional Components of Cauliflower Curd

Constituents	Quantity	Constituents	Quantity
Moisture (g/100 g)	90.8	Potassium (mg/100 g)	320
Carbohydrates (g/100 g)	4.0	Energy (Kcal/100g)	30
Proteins (g/100 g)	2.6	Vitamin K	16.59
Fats (g/100 g)	0.4	Thaimine (mg/100 g)	0.04
Dietary Fibre (g/100 g)	2.1	Riboflavin (mg/100 g)	0.10
Minerals (g/100 g)	1.0	Niacin (mg/100 g)	1.0
Calcium (mg/100 g)	33	Pantothenic acid(mg/100 g)	0.71
Phosphorus (mg/100 g)	57	Vitamin B6 (mg/100 g)	0.20
Iron (mg/100 g)	1.5	Folic acid (mg/100 g)	61
Magnesium (mg/100 g)	16.1	Vitamin C (mg/100 g)	56

Goplan *et al.*, 1985)

Phytochemicals

Raw cauliflower is rich in vitamins C, K, B-6 and folate. Vitamin C helps to strengthen the immune system and aids in collagen production. Vitamin K is essential for strong bones and proper blood clotting. Vitamin B-6 is necessary for producing red blood cells, for energy metabolism and helps to manufacture brain chemicals, or neurotransmitters. Folate enhances immune system function and ensures a healthy pregnancy. Cauliflower being rich in these nutrients, its consumption contributes to better health.

Cauliflower also contains the minerals potassium and manganese. Potassium, a mineral found in many fruits and vegetables, is needed to regulate heartbeat and blood pressure. Manganese, a trace mineral, is important for maintaining healthy, strong bones as well as the metabolism of proteins, carbohydrate and cholesterol.

Cauliflower contains two phytonutrients or chemicals found in plants known as sulforaphane and isothiocyanates, which neutralize carcinogens and can prevent esophageal and lung cancer, as well as reduce the risk of other types of cancer. It also contains indole-3-carbinol (IC3). They increase the body's production of certain enzymes that sweep toxins and carcinogens out of the system and prevent other enzymes from activating cancer-causing agents in the body before they can damage cells. Also, IC3 is a powerful anti-tumor agent, particularly beneficial for preventing tumor growth in breast, cervical, prostate cells and is also anti-inflammatory.

Cauliflower also contains a compound called glucoraphin, which protects the stomach and intestines from cancer and ulcers. Researchers have determined that the sulforaphane made from a glucosinolate in cauliflower (glucoraphanin) can help protect the lining of stomach. Sulforaphane provides health benefit by preventing bacterial overgrowth of *Helicobacter pylori* in the stomach or too much clinging by this bacterium to the stomach wall.

Cauliflower is healthful, filling and low in calories. It is also a good food source of dietary fiber. These qualities make it a good choice for losing weight, preventing weight gain and maintaining regularity. However, according to the World's Healthiest Foods, cauliflower contains naturally occurring substances found in many plant and animal foods called purines which is related with gout.

Health Benefits of Cauliflower based on Traditional Knowledge/Folklore practices

Cancer Prevention: A diet high in cruciferous vegetables such as cauliflower has been been linked to a significant reduction in the risk of cancer, especially prostate cancer, breast cancer, colon cancer, ovarian cancer, and bladder cancer. One Canadian study found that eating a half cup of cauliflower per day reduced the risk of prostate cancer by 52%.

Digestive Support: Cauliflower is a great source of dietary fiber, which is essential for easy digestion and complete evacuation.

Anti-Inflammatory: Cauliflower is good for preventing chronic inflammatory aphanina like arthritis and gout.

Pregnancy: Cauliflower provides a good amount of folate (B9), a B vitamin that is necessary for a healthy pregnancy. Folate deficiency in pregnant women can lead to problems such as birth defects and low birth weight. In addition to folate, cauliflower is also loaded with other important B vitamins like niacin, riboflavin, pantothenic acid, and thiamine.

Cardiovascular Support

Cauliflower contains a high number of antioxidants, which are essential for the body's overall health and help to prevent heart disease and stroke. Antioxidants are also essential in destroying free radicals that accelerate the signs of aging. The anti-inflammatory support provided by cauliflower (including its vitamin K and omega-3 content) makes the food also capable of providing cardiovascular benefits. Glucoraphanin is a glucosinolate that can be converted into the isothiocyanate (ITC) sulforaphane. The sulforaphane trigger anti-inflammatory activity in cardiovascular system which probably help reverse blood vessel damage. Cauliflower also contains allicin, which has been found to reduce the occurrence of stroke and heart disease. Additionally, cauliflower can help to lower cholesterol levels in the body.

Therapeutic properties of Cauliflower

One of the major health benefits of cauliflower is its cancer preventing property. Steinkellner *et al.*, (2001) studied the impact of cruciferous vegetables and their constituents on enzymes that are involved in the metabolism of DNA-reactive carcinogens. Brassica juices induced glutathione-S-transferases (GST) and cytochrome P-450 1A2 in human hepatoma cells (HepG2) that protect against the genotoxic effects of B(a)P and other carcinogens. They found that the isoenzyme induced was GST-pi which plays an important role in protection against breast, bladder, colon and testicular cancer.

Sulforaphane present in cruciferous vegetables plays an important role in prevention of cancer. At the Johns Hopkins University in Baltimore, Maryland, 69 percent of the rats injected with a chemical known to cause mammary cancer developed tumors while only 26 percent of the rats given the carcinogenic chemical plus sulforaphane showed cancer.

There are three systems in the body (1) the body's detox system, (2) its antioxidant system, and (3) its inflammatory/anti-inflammatory system. Chronic imbalances in any of these three systems can increase risk of cancer, and when imbalances in all three systems occur simultaneously, the risk of cancer increases significantly. The phytochemicals maintain this balance through the ingrediens present in them and prevent the occurrence of these diseases.

The detox support provided by cauliflower includes antioxidant nutrients to boost Phase 1 detoxification activities and sulfur-containing nutrients to boost Phase 2 activities. Cauliflower also contains phytonutrients called glucosinolates that can help activate detoxification enzymes and regulate their activity. Three glucosinolates that have been clearly identified in cauliflower are glucobrassicin, glucoraphanin, and gluconasturtiian. While the glucosinolate content of cauliflower is definitely significant from a health standpoint, cauliflower contains about one-fourth as much total glucosinolates as Brussels sprouts, about one-half as much as Savoy cabbage, about 60% as much as broccoli, and about 70% as much as kale.

This broad-spectrum antioxidant support helps lower the risk of oxidative stress in our cells. Chronic oxidative stress – meaning chronic presence over overly reactive oxygen-containing molecules and cumulative damage to the body cells by these molecules – is a risk factor for development of most cancer types. By providing with such a great array of antioxidant nutrients, cauliflower helps lower cancer risk by helping to avoid chronic and unwanted oxidative stress.

The health benefits associated to the antioxidant properties present in cauliflower reinforce their contribution to a healthy and balanced diet (Batisa, 2011). Horst *et al.*, (2010) were of the opinion that the Brassica vegetables presented protection against DNA damage, an effect possibly related to increased hepatic lutein concentrations.

Cruciferous vegetables, including cauliflower, contain goitrin, thiocyanate, and isothiocyanate. These chemicals, known collectively as goitrogens, inhibit the formation of thyroid hormones and cause the thyroid to enlarge in an attempt to produce more. Goitrogens are not hazardous for healthy people who eat moderate amounts of cruciferous vegetables, but they may pose problems for people who have a thyroid condition or are taking thyroid medication.

References

Batista, C., Barros, L., Carvalho, A.M., and Ferreira, I.C. 2011. Nutritional and nutraceutical potential of rape (*Brassica napus* L. var. *napus*) and "tronchuda" cabbage (*Brassica oleraceae* L. var. *costata*) inflorescences. Food Chem Toxicol., (2011), 49(6):1208-14.

Gopalan, C., Sastri, B.V.R. and Balasubramanian, S.C. 1985. Nutritive Value of Indian Foods. National Institue of Nutrition, ICMR, Hyderabad, India.

Horst, M.A., Ong, T.P., Jordão, A.A. Jr., Vannucchi, H., Moreno, F.S., and Lajolo, F.M. 2010. Water extracts of cabbage and kale inhibit ex vivo H(2)O(2)-induced DNA damage but not rat hepatocarcinogenesis. Braz J Med Biol Res., (2010), 43(3):242-8.

Steinkellner, H., Rabot, S., Freywald, C., Nobis, E., Scharf, G., Chabicovsky, M., Knasmüller, S., and Kassie, F. 2001. Effects of cruciferous vegetables and their constituents on drug metabolizing enzymes involved in the bioactivation of DNA-reactive dietary carcinogens. Mutation Research/Molecular Mechanisms of Anticarcinogenesis and Antimutagenesis. Fundamental and Molecular Mechanisms of Mutagenesis, 480-481: 285-297.

3.3. Broccoli (*Brassica oleracea* L. var. *italica* Plenck)

Sprouting Braccoli is sometimes called as broccoli and is a winter cauliflower. Braccoli is a member of the Brassicaceae as all other members of this family like cabbage, cauliflower, kale, *etc*. The name broccoli refers to young shoots which develop in spring on some species of genus Brassica ('brocco' in Italian means a shoot). Its scientific name is *Brassica oleracea L.* and in USA it is also known as 'Italian Broccoli'. Morphologically, sprouting broccoli resembles cauliflower, but its terminal head is rather loose, green in colour and the flower stalks are longer than cauliflower. The sprouts in the axils of leaves develop after removal of the terminal head. Both terminal head and the sprouts are consumed as vegetable.

Cultivars/Varieties: There are two variants in this, green and purple. The green sprouting broccoli is classified according to its maturity. The early ones' in this group are De Cicco, Green Bud and Spartan Early. The late varieties are Walthan 29, Green Mountain, Coastal and Atlantic. The F1 hybrids are Southern Comet, Premium Crop, Clipper, Laser, Corsair, Excalibur, Cruiser, Emerald Corona and Late Corona, Stiff, Kayak and Green Surf. Some of the green varieties released in India are Pusa broccoli Kt, Sel-1, Palam Samridhi (DPGB-1) and Punjab broccoli.

Uses

Its main usage is as vegetable.

Nutritional components of Broccoli

Constituents	Quantity (per 100g)	Constituents	Quantity (per 100g)
Moisture	89.9 g	Potassium	3.52 mg
Carbohydrates	5.5 g	Copper	0.8 mg
Proteins	3.3 g	Calcium	1.2 mg
Fats	0.2 g	Phosphorus	0.79 mg
Dietary Fibre	2.37	Vitamin A	9000 IU
Energy	37 Kcal	Thaimine	33 IU
Iron	20.5 mg	Vitamin C	137 IU
Sulphur	1.26 mg	Vitamin K	92.46 mcg

(Chatterjee, 1986)

Broccoli is an excellent source of immune-supportive vitamin C, anti-inflammatory vitamin K, and heart-healthy folate. It is a very good source of free-radical-scavenging vitamin A (through its concentration of carotenoid phytonutrients), enzyme-activating manganese and molybdenum; digestive-health-supporting fiber; heart-healthy potassium and vitamin B6; and energy-producing vitamin B2 and phosphorus. It is a good source of energy-producing vitamin B1, vitamin B3, vitamin B5, protein, and iron; bone-healthy magnesium and calcium; and antioxidant-supportive vitamin E and selenium. Besides these nutrients' broccoli has several other phytochemicals which prevent many diseases.

3. Cole Crops 63

Phytochemicals

Broccoli is a particularly rich source of a flavonoid called kaempferol, which has shown its ability to lessen the impact of allergy-related substances in the body. Glucoraphanin, gluconasturtiian, and glucobrassicin are 3 glucosinolate phytonutrients found in a special combination in broccoli. Broccoli provides many flavonoids in significant amounts, including the flavonoids kaempferol and quercitin. Also concentrated in broccoli are the carotenoids lutein, zeaxanthin, and beta-carotene. All three of these carotenoids function as key antioxidants. Isothiocyanates (ITCs) made from the glucosinolates in broccoli are modifiers of the first step in detoxification (called Phase I).

Therapeutic Properties of Braccoli

Various epidemiologic studies have indicated that consumption of broccoli is associated with a lower risk of cancer (Clarke *et al.*, 2008), including breast (Ambrosone *et al.*, 2004), prostate (Joseph *et al.*, 2004), lung, stomach (van *et al.*, 1999), and colon cancers (Chung *et al.*, 2000). The anticancer effect of broccoli has been attributed to sulforaphane (SFN), an isothiocyanate formed by hydrolysis of a precursor glucosinolate called "glucoraphanin" (Clarke *et al.*, 2008). Although glucoraphanin is found in varying amounts in all cruciferous vegetables, the highest concentration of this compound is found in broccoli and its sprouts (Zhang *et al.*, 1992). Among various parts of mature broccoli, the florets have the maximum amount of SFN.The amount of SFN in 1 g of dry broccoli florets ranges from 507 to 684 µg (Campas-Baypoli *et al.*, 2010). The sprouts of broccoli seem to have 20 to 30 times higher concentration of glucoraphanin, an SFN precursor (Fahey *et al.*, 1997; Zhang, 1994). Glucoraphanin is converted to SFN by myrosinase, an enzyme released from broccoli during its consumption and also found in our stomach (Clarke *et al.*, 2008). The reduced risk of cancer after consumption of broccoli is associated with the ability of SFN to inhibit phase 1 enzymes (implicated in the conversion of pro-carcinogens to carcinogens) and induce phase 2 enzymes, that are implicated in detoxification and excretion of carcinogens from body (Clarke *et al.*, 2008; Zhang *et al.*, 1992). SFN has also been shown to induce apoptosis, or programmed cell death, in cancer cells. Treatment with 15 µM SFN was found to induce apoptosis in both the p53-positive and p53-negative human colon cancer cell lines (Pappa *et al.*, 2006).

When threatened with dangerous levels of potential toxins, or dangerous numbers reactive oxygen species molecules, signals are sent within the body to the inflammatory system, directing it to "kick in" and help protect the body from potential damage. One key signaling device is a molecule called "**NF-kappaB**". The anti-inflamatory benefit of broccoli has been related with NF-kappaB signaling system. Research studies have made it clear that the NF-kappaB signaling system that is used to "rev up" our inflammatory response can be significantly suppressed by isothiocyanates (ITCs). Broccoli has yet another anti-inflammatory trick, because it is a rich source of one particular phytonutrient (a flavonol) called kaempferol. Especially inside of our digestive tract, kaempferol has the ability to lessen the impact of allergy-related substances (by lowering the immune system's production of IgE-antibodies).

By lessening the impact of allergy-related substances, the kaempferol in broccoli can help lower the risk of chronic inflammation. Considered as a group, the vitamins, minerals, flavonoids, and carotenoids contained in broccoli work to lower risk of oxidative stress in the body.

References

Ambrosone, C.B., McCann, S.E., Freudenheim, J.L., Marshall, J.R., Zhang, Y., and Shields, P.G 2004. Breast cancer risk in premenopausal women is inversely associated with consumption of broccoli, a source of isothiocyanates, but is not modified by GST genotype. J Nutr., 134: 1134–1138.

Chatterjee, S.S. 1986. In: Vegetable crops in India. Cole Crops. (Eds. Bose, T.K and Som, M.G), Naya Prokash Publishers, Calcutta. Pp.168-170.

Campas-Baypoli, O,N., Sanchez-Machado, D.I., Bueno-Solano, C., Ramirez-Wong, B., and Lopez-Cervantes, J. 2010. HPLC method validation for measurement of sulforaphane level in broccoli by-products. Biomed Chromatogr., 24: 387–392.

Chung, F.L., Conaway, C.C., Rao, C.V., and Reddy, B.S. 2000. Chemoprevention of colonic aberrant crypt foci in Fischer rats by sulforaphane and phenethyl isothiocyanate. Carcinogenesis, 21: 2287–2291.

Clarke, J.D., Dashwood, R.H., and Ho, E. 2008. Multi-targeted prevention of cancer by sulforaphane. Cancer Lett., 269: 291–304.

Fahey, J.W., Zhang, Y., and Talalay, P. 1997. Broccoli sprouts: an exceptionally rich source of inducers of enzymes that protect against chemical carcinogens. Proc Natl Acad Sci., USA, 94: 10367–10372.

Joseph, M.A., Moysich, K.B., Freudenheim, J.L., Shields, P.G., Bowman, E.D., Zhang, Y., Marshall, J.R., and Ambrosone, C.B. 2004. Cruciferous vegetables, genetic polymorphisms in glutathione S-transferases M1 and T1, and prostate cancer risk. Nutr Cancer, 50: 206–213.

Pappa, G., Lichtenberg, M., Iori, R., Barillari, J., Bartsch, H., and Gerhauser, C. 2006. Comparison of growth inhibition profiles and mechanisms of apoptosis induction in human colon cancer cell lines by isothiocyanates and indoles from Brassicaceae. Mutat Res., 599: 76–87.

van Poppel, G., Verhoeven, D.T., Verhagen, H., and Goldbohm, R.A. 1999. Brassica vegetables and cancer prevention. Epidemiology and mechanisms. Adv Exp Med Biol., 472: 159–168.

Zhang, Y., Talalay, P., Cho, C.G., and Posner, G.H. 1992. A major inducer of anticarcinogenic protective enzymes from broccoli: isolation and elucidation of structure. Proc Natl Acad Sci., USA, 89: 2399–2403.

Zhang, Y., and Talalay, P. 1994. Anticarcinogenic activities of organic isothiocyanates: chemistry and mechanisms. Cancer Res., 54: 1976s–1981s.

3.4. Brussels Sprouts (*Brassica oleracea* L. var. *gemmifera* Zenk.)

Brussel sprouts are a type of non-heading cauliflower. Brussel sprouts are an herbaceous biennial that lack an apical head but have axillary heads or sprouts that are produced along an elongated stem. This crop is of recent development in cole group and as the name implies, the crop was initially grown mainly in the vicinity of Brussel in Belgium. In India it is grown in a very limited measure mainly for supply to hotels in cosmopolitan cities.

Cultivars/Varieties: In Brussel Sprouts, there are two types, dwarf and tall cultivars. The dwarf varieties are short stemmed, mostly less than 50 cm in length. They are early cropper and the popular cultivars are Early Morn, Dwarf Improved, Fruher Zwerg and Kvik. Jade Cross is a F1 hybrid. The tall cultivars are preferred for longer seasons and are classified according to their maturity and the size of sprouts. The early varieties in this are Evesham and Bedfordshire.

Uses

Brussel Sprouts are mainly used as vegetable in star hotels and exotic cuisine. It is mainly grown in temperate reiongs and supplied to urban markets.

Nutritional Components in Brussel Sprouts

All cruciferous vegetables provide integrated nourishment across a wide variety of nutritional categories and provide broad support across a wide variety of body systems as well. Brussels sprouts are members of the Brassica family and therefore akin to broccoli and cabbage. Brussels sprouts are rich in many valuable nutrients. They are an excellent source of vitamin C and vitamin K. They are a very good source of numerous nutrients including folate, vitamin A, manganese, dietary fiber, potassium, vitamin B6 and thiamin (vitamin B1) and a good source of omega-3 fatty acids, iron, phosphorus, protein, molybdenum, magnesium, riboflavin (vitamin B2), vitamin E, calcium, and niacin.

Phytochemicals

In addition to these nutritional components cited above, Brussels sprouts contain numerous disease-fighting phytochemicals including sulforaphane, indoles, glucosinolates, isothiocynates, coumarins, dithiolthiones, and phenols. Flavonoid antioxidants like isorhamnetin, quercitin, and kaempferol are also found in Brussels sprouts, that serve as the antioxidants.

Jaiswal *et al.*, (1995) studied the phenolic composition, antibacterial activity and antioxidant capacity of selected Brassica vegetables, including York cabbage, Brussels sprouts, broccoli and white cabbage after extraction with aqueous methanol as a solvent. HPLC-DAD analysis showed that different vegetables contain a mixture of distinct groups of phenolic compounds. All the extracts studied showed a rapid and concentration dependent antioxidant capacity in diverse antioxidant systems. York cabbage extract exhibited significantly higher antibacterial activity against *Listeria monocytogenes* (100%) and *Salmonella abony* (94.3%), being the most susceptible at a concentration of 2.8%, whereas broccoli, Brussels sprouts and white cabbage had moderate to weak activity against all the test organisms.

Studies by Cartea *et al.*, (2010) indicated that among the various types of phenolic compounds like phenolic acids, hydroxycinnamic acid derivatives and flavonoids, had some of the beneficial effects on human health and the influence of environmental conditions and processing mechanisms determind its efficacy.

As Brussel Sprouts is relatively a new crop, not much of trational knowledge on its use in folklore is available.

Research findings supporting the health benefits of Brussel Sprouts

Cancer Prevention: Among all types of cancer, prevention of the bladder cancer, breast cancer, colon cancer, lung cancer, prostate cancer, and ovarian cancer are most closely associated with intake of Brussels sprouts. These properties are believed to be due to glucosinolates present in Brussels sprouts and their detox-activating isothiocyanates. Their details are listed below:

Glucosinolate	Derived Isothiocyanate	Isothiocyanate Abbreviation
glucoraphanin	sulforaphane	SFN
glucobrassicin	indole-3-carbinol*	I3C (ITC)
sinigrin	allyl-isothiocyanate	AITC
gluconasturtiian	phenethyl-isothiocyanate	PEITC

Nijhoff *et al.*, (1995) tried to study possible modulating effects of consumption of Brussels sprouts on duodenal, rectal and lymphocytic (i) glutathione S-transferase (GST) enzyme activity, (ii) GST isozyme levels and (iii) glutathione (GSH) content. They found that consumption of glucosinolate-containing Brussels sprouts for 1 week resulted in increased rectal GST-alpha and -pi isozyme levels. They hypothesized that these enhanced detoxification enzyme levels may partly explain the epidemiological association between a high intake of glucosinolates (cruciferous vegetables) and a decreased risk of colorectal cancer.

Removal of toxins from the body: Brussel sprouts contain a sulfur-containing compound called D3T. (D3T is the abbreviated name for 3H-1,2-dithiole-3-thione). Though its mechanism of action is not fully understood, it is believed to play a role in detoxifying the toxins from the body.

Antioxidant activity: Flavonoid like isorhamnetin, quercitin, and kaempferol found in Brussels sprouts, serve as the antioxidants. A host of other antioxidant ingredients found in Brussels sprouts, include Vitamins C, E, and A, as well as the mineral manganese provide protection against oxidative stress on the body's cells.

Anti-inflamatory activity: Brussels sprouts help to regulate the body's inflammatory/ anti-inflammatory system and prevent unwanted inflammation. Particularly well-studied in this context is the glucosinolate called glucobrassicin. The glucobrassicin found in Brussels sprouts can get converted into an isothiocyanate molecule called ITC, or indole-3-carbinol or I3C. I3C is an anti-inflammatory compound that can actually operate at the genetic level, and by doing so, prevent the initiation of inflammatory responses at a very early stage.

Not only does this ITC trigger anti-inflammatory activity in our cardiovascular system & mash; it may also be able to help prevent blood vessel damage.

Digestive System Protection: The fibre content present in Brussel Sprouts help in maintaining a better digestive health system.

Bone Health: Brussels sprouts are especially high in vitamin K (one cup contains 273.5% of the RDA), which promotes healthy bones, prevents calcification of the body's tissues.

References

Ambrosone, C.B., and Tang, L. 2009. Cruciferous vegetable intake and cancer prevention: role of nutrigenetics. Cancer Prev Res (Phila Pa)., (2009), 2(4):298-300.

Angeloni, C., Leoncini, E., Malaguti, M., Angelina, S., Hrelia, P., and Hrelia, S. 2009. Modulation of phase II enzymes by sulforaphane: implications for its cardioprotective potential. J Agric Food Chem., (2009), 57(12):5615-22.

Antosiewicz, J., Ziolkowski, W., Kar, S., Powlony, A.A., and Singh, S.V. 2008. Role of reactive oxygen intermediates in cellular responses to dietary cancer chemopreventive agents. Planta Med., (2008), 74(13):1570-9.

Carpenter, C.L., Yu, M.C., and London, S.J. 2009. Dietary isothiocyanates, glutathione S-transferase M1 (GSTM1), and lung cancer risk in African Americans and Caucasians from Los Angeles County, California. Nutr Cancer., (2009), 61(4):492-9.

Cartea, M.E., Francisco, M., Soengas, P., and Velasco, P. 2010. Phenolic compounds in Brassica vegetables. Molecules, (2010), 16(1):251-80.

Bryant, C. B., Sanjeez, K., Sreedhar, C., Shah, J., Pal, J., Haider, M., Seward, S., Qazi, A.M., Morris, R., Seaman, A., Shammas, M.A., Steffen, C., Patti, R.B., Prasad, M., Weaver, D.W., and Batchu, R.B. 2010. Sulforaphane induces cell cycle arrest by protecting RB-E2F-1 complex in epithelial ovarian cancer cells. Molecular Cancer, (2010), 9(1): 47.

Clarke, J.D., Dashwood, R.H., and Ho, E. 2008. Multi-targeted prevention of cancer by sulforaphane. Cancer Lett., (2008), 269(2):291-304.

Jaiswal, A.K., Rajauria, G., Abu-Ghannam, N., and Gupta, S. 2011. Phenolic composition, antioxidant capacity and antibacterial activity of selected Irish Brassica vegetables. Nat Prod Commun., (2011), 6(9):1299-304.

Nijhoff, W.A., Grubben, M.J., Nagengast, F.M., Jansen, J.B., Verhagen, H., van Poppel, G., and Peters, W.H. 1995. Effects of consumption of Brussels sprouts on intestinal and lymphocytic glutathione S-transferases in humans. Carcinogenesis. (1995), 16(9):2125-8.

4. Solanaceous Vegetables

In Solanaceae family there are 85 genera, which includes both tuberiferous and non-tuberiferous plants. Among the non-tuberiferous, tomato, brinjal and chilli are the important fruit vegetables.

4.1. Tomato (*Lycopersicon esculentum* Mill.)

Tomato has originated in Peru and Mexican region of South America. It got its name form the azetc work "Tomatl". In India tomato is the second largest cultivated vegetable after potato. Though it is produced in almost entire country, the major producer states are Andhra Pradesh, Karnataka, Orissa, West Bengal, Bihar, Gujarat, Madhya Pradesh and Maharashtra.

Cultivar/Varieties: A large number of varieties/hybrids of tomatoes are under cultivation. The most popular among them are Arka Rakshak, Arka Samrat, Arka Saurabh, Arka Vikash, ARTH 3, ARTH 4, Avinash 2, BSS 90, CO3, HS 101, HS 102, HS 110. Hisar Anmol, Hisar Arun, Hisar Lalima, Hisar Lalit, Krishna, KS 2, Matri, MTH 6, NA 601, Naveen, Pusa 120, Punjab Chhuhara, Pant Bahar, Pusa Divya, Pusa Early Dwarf, Pusa Hybrid 1, Pusa Hybrid 2, Pusa Hybrid 4, Pusa Ruby, Pusa Sheetal, Pusa Uphar, Rajni, Rashmi, Ratna and Rupali.

Uses

Tomato adds richness to any dish it is added. Besides using for culinary purpose tomato is also used as salad and dressings and also as adjuvant in several locally available fresh snacks. It is used for thickening of gravies in many Indian curries. As a processed product tomato sauce and ketchup are on top of favourites. It is not only delicious in taste but also rich in its medicinal value. The pulp and juice are mild aperient, a promoter of gastric secretion and blood purifier. It is also considered to be intestinal antiseptic. It is said to be useful in sour mouth, canker of mouth and chronic dyspepsia.

Nutritional Components of Tomato

Constituent	Quantity/100g	Constituent	Quantity/100g
Moisture (g)	93.1	Manganese (mg)	0.12
Protein (g)	1.9	Iron (mg)	1.8
Carbohydrates (g)	3.6	Sulphur (mg)	24
Fats (g)	0.1	Potassium (mg)	114
Fibre (g)	0.7	Carotene (µg)	192
Minerals (g)	0.6	Thiamime (mg)	0.07
Calcium (mg)	20	Riboflavin (mg)	0.01
Phosphorus (mg)	36	Niacin (mg)	0.4
Magnesium (mg)	15	Vitamin C (mg)	31
Copper (mg)	0.06	Energy (Kcal)	23

(Tiwari and Choudhury, 1986; Gopolan *et al.*, 1985)

Phytochemicals

Tomato is rich in lycopene (a carotenoid phytonutrient widely recognized for its antioxidant properties) and provide a unique variety of phytonutrients. These include additional carotenoids (including beta-carotene, lutein, and zeaxanthin); flavonoids (including naringenin, chalconaringenin, rutin, kaempferol, and quercetin); hydroxycinnamic acids (including caffeic, ferulic, and coumaric acid); glycosides (including esculeoside A); and fatty acid derivatives (including 9-oxo-octadecadienoic acid).

- Tomatoes are also rich in several phytonutrients, which include:
- Flavonones – naringenin & chalconaringenin
- Flavonols – rutin, kaempferol & quercetin
- Hydroxycinnamic acids -caffeic acid, ferulic acid, coumaric acid
- Carotenoids – lycopene, lutein
- Glycosides – esculeoside A
- Fatty acid derivatives – 9-oxo-octadecadienoic acid

Tomatoes are also an excellent source of vitamin C and vitamin A as well as bone-healthy vitamin K. They are a very good source of enzyme-promoting molybdenum; heart-health protecting potassium and magnesium, vitamin B6, folate, and dietary fiber; blood sugar-balancing manganese. In addition, tomatoes are a good source of niacin, and vitamin E; energy-producing iron, vitamin B1, and phosphorus; muscle-building protein, and bone-health promoting copper. Two little-known phytonutrients – one called esculeoside A and the other called 9-oxo-octadecadienoic acid – are currently under active investigation by researchers as tomato phytonutrients especially important in blood fat regulation.

Health Benefits of Tomato

Tomato intake has been shown to result in decreased total cholesterol, decreased LDL cholesterol, and decreased triglyceride levels. It's also been shown to decrease accumulation of cholesterol molecules inside of macrophage cells.

Risk for many cancer types starts out with chronic oxidative stress and chronic unwanted inflammation. Alpha-tomatine a saponin phytonutrient present in tomato has shown its ability to alter metabolic activity in developing prostate cancer cells. There is fairly well documented risk reduction for breast cancer in association with lycopene intake. Several other health benefits of consuming tomatoes have been indicated in literature.

Antioxidant: Tomatoes contain a lot of vitamins A and C, mostly because of beta-carotene, and these vitamins act as an antioxidant, working to neutralize dangerous free radicals in the blood stream.

Diabetes: Tomatoes also have plenty of the mineral chromium, which helps diabetics to keep their blood sugar level under control.

Smoking: Tomaotes may help reduce the damage due to smoking. Tomatoes contain chlorogenic acid and coumaric acid, which help to fight against some of the carcinogens brought about by cigarette smoke.

Vision: Because of all that vitamin A, tomatoes are also an excellent food to help improve the vision. This also means tomatoes can help to reduce night blindness.

Heart diseases: Due to potassium and vitamin B, tomatoes help to lower blood pressure and to lower high cholesterol levels. This, in turn, could help prevent strokes, heart attack and other potentially life-threatening heart problems.

Skin care: Because of high amounts of lycopene, a substance found in many of the more expensive over-the-counter facial cleansers, tomatoes are may be good for skin care.

Cancer: Various studies have shown that because of lycopene in tomatoes, the red fruit helps to lessen the chances of prostate cancer in men, and also reduces the chance of stomach cancer and colorectal cancer. Lycopene is highly potent antioxidant that may help to stop the growth of cancer cells.

Bones: Tomatoes have a fair amount of vitamin K and calcium, both of which help to strengthen and possibly repair in minor ways bones and bone tissue.

Kidney stones and gallstones: Eating tomatoes without the seeds has been shown in some studies to lessen the risk of gallstones and kidney stones.

Therapeutic properties of tomato

International epidemiological research has provided information on risk factors and preventive approaches in chronic diseases. Individuals in the Mediterranean area were found to have a lower risk of several important chronic diseases, including coronary heart disease and a number of types of cancer such as breast, colon, and prostate cancer. Vegetables and fruits in general and cooked tomatoes, together with olive oil, was reported to lower the risk. These results lead to public health recommendations to consume more vegetables and, especially, cooked tomatoes with olive oil (Weisburger, 2002).

Many of the nutrients present in tomatoes may function individually, or in concert, to protect lipoproteins and vascular cells from oxidation, the most widely accepted theory for the genesis of atherosclerosis. This hypothesis has been supported by *in-vitro*, limited *in-vivo*, and many epidemiological studies that associated reduced cardiovascular risk with consumption of antioxidant-rich foods like tomatoes. Other cardioprotective functions provided by the nutrients in tomatoes include the reduction of low-density lipoprotein (LDL) cholesterol, homocysteine, platelet aggregation, and blood pressure (Willcox et al., 2003). Ried and Fakler (2011), based on their studies, suggested that lycopene taken in doses ≥ 25mg daily is effective in reducing LDL cholesterol by about 10% which is comparable to the effect of low doses of statins in patient with slightly elevated cholesterol levels. However, they opined that more research is needed to confirm suggested beneficial effects on total serum cholesterol and systolic blood pressure.

Though majority of evidence came from observational studies, recent human clinical trials and animal studies have provided additional support to the belief that tomato nutrients helps in prevention of prostate cancer (Fraser *et al.*, 2005). Most of the clinical trials with tomato products also suggest a synergistic action of lycopene with other nutrients, in lowering biomarkers of oxidative stress and carcinogenesis (Basu, 2007). Erdman *et al.*, (2009) postulated that metabolic products of lycopene, the lycopenoids, may be responsible for some of lycopene's reported bioactivity. Some believe that the mechanism underlying the inhibitory effects of lycopene on carcinogenesis could involve ROS scavenging, up-regulation of detoxification systems, interference with cell proliferation, induction of gap-junctional communication, inhibition of cell cycle progression and modulation of signal transduction pathways (Bhuvanewari and Nagini, 2005). A clear understanding of the molecular mechanisms of action of lycopene is crucial in the valuation of this molecule as a potential preventive and therapeutic agent (Palozza *et al.*, 2010).

However, the USFDA found no credible evidence for an association between tomato consumption and a reduced risk of lung, colorectal, breast, cervical, or endometrial cancer (Kavanaugh *et al.*, 2007). But some researchers are of the opinion that much more complementatry research in the field of nutrigenomics have to be done to understand the nutritional benefits better from the human side and the role and mechanism of lycopene in disease prevention (Hall *et al.*, 2008; Story *et al.*, 2010).

References

Basu, A., and Imrhan, V. 2007. Tomatoes versus lycopene in oxidative stress and carcinogenesis: conclusions from clinical trials. Eur J Clin Nutr., (2007), 61(3):295-303.

Bhuvaneswari, V., and Nagini, S. Lycopene: 2005.a review of its potential as an anticancer agent. Curr Med Chem Anticancer Agents, (2005), 5(6):627-35.

Erdman, J.W. Jr., Ford, N.A., and Lindshield, B.L. 2009. Are the health attributes of lycopene related to its antioxidant function? Arch Biochem Biophys., (2009), 483(2):229-35.

Fraser, M.L., Lee, A.H., and Binns, C.W. 2005.Lycopene and prostate cancer: emerging evidence. Expert Rev Anticancer Ther., (2005), 5(5):847-54.

Gopalan, C., Sastri, B.V.R., and Balasubramanian, S.C. 1985. Nutritive Value of Indian Foods. National Institue of Nutrition, ICMR, Hyderabad, India.

Hall, R.D., Brouwer, I.D., and Fitzgerald, M.A. 2008. Plant metabolomics and its potential application for human nutrition. Physiol Plant., (2008), 132(2):162-75.

Kavanaugh, C.J., Trumbo, P.R., and Ellwood, K.C. 2007. The U.S. Food and Drug Administration's evidence-based review for qualified health claims: tomatoes, lycopene, and cancer. J Natl Cancer Inst., (2007), 18;99(14):1074-85.

Palozza, P., Parrone, N., Catalano, A., and Simone, R. 2010. Tomato lycopene and inflammatory cascade: basic interactions and clinical implications. Curr Med Chem., (2010), 17(23):2547-63.

Ried, K., and Fakler, P. 2011.Protective effect of lycopene on serum cholesterol and blood pressure: Meta-analyses of intervention trials. Maturitas, (2011), 68(4):299-310.

Story, E.N., Kopec, R.E., Schwartz, S.J., and Harris, G.K. 2010. An update on the health effects of tomato lycopene. Annu Rev Food Sci Technol., (2010), 1:189-210.

Tiwari, R.N., and Choudhury, B. 1986. In: Vegetbles crops in India. Solanaceous Crops: Tomato. (eds. Bose, T.K. and Som, M.G), Nayaprokash Publisher, Culcutta. Pp.248-249.

Weisburger, J.H. 2002.Lycopene and tomato products in health promotion. Exp Biol Med (Maywood), (2002), 227(10):924-7.

Willcox, J.K, Catignani, G.L., and Lazarus, S. 2003.Tomatoes and cardiovascular health. Crit Rev Food Sci Nutr., (2003), 43(1):1-18.

4.2. Eggplant or Brinjal (*Solanum melongena*, L)

Eggplant or brinjal belongs to the Solanaceae or nightshade family. Egg plant is so named because early cultivars were egg-shaped fruits. However, present varieties vary in shape from oval to round and long to ablong. The colour of mature fruit is purple to purple black, but can also be red, yellowish white, white or green. Though brinjal originated in Indo-Burma region, it is distributed in South, South East Asia, Southern Europe, China, Japan and America too. Brinjal is one of the most common tropical vegetables grown in India. It is known by different names in different states.

Varieties: Based on maturity the early varieties/hybrids widely grown are ABH 1, ABH 2, Arka Navneet, Arka Neelkantha, Arka Nidhi, Azad Kranti, Mysore Green, NDBH 1, Pant Samrat, Punjab Chamkila, Punjab Sadabahar, Pusa Bhairav, Pusa Bindu, Pusa Hybrid 5, Pusa Hybrid 6, Pusa Hybrid 9, Pusa Upkar, Pusa Uttam and Pusa Purple Long (Extra early). Medium maturity ones' are Annamalai, Arka Keshav, Arka Kusumakar, Arka Sheel, Arka Shirish, Azad B 1, BR 112, Co 1, Gujarat 6, Hisar Jamuni, Hisar Shyamal, Jamuni Gole, Junagarh Long, MDU 1, Pant Rituraj, PH 4, Punjab Barsati, Punjab Neelam, Pusa Anupam, Pusa Kranti, Pusa Purple Cluster and Vaishali. The late maturity varieties are Manjri Gota and Punjab Bahar.

Uses

Eggplant fruit is used in different ways, as baked, sautéed, cut into strips or cubes and fried, or stuffed. In northern parts of India, it is baked and crushed with spices.

Nutritional Components of Eggplant fruit (Brinjal)

Constituent	Quantity/100g	Constituent	Quantity/100g
Moisture	92.7 g	Manganese	0.20 mg
Protein	1.4 g	Iron	0.90 mg
Carbohydrates	4.0 g	Potassium	188 mg
Fats	0.3 g	Carotene	74 µg
Fibre	1.3 g	Thiamime	0.04 mg
Minerals	0.3 g	Riboflavin	0.11 mg
Calcium	18 mg	Niacin	0.90 mg
Phosphorus	47 mg	Vitamin C	12 mg
Magnesium	11.4 mg	Energy (Kcal)	24 Kcal
Copper	0.07 mg		

Source: (Gopalan *et al.*, 1985)

Eggplant is an excellent source of digestion-supporting dietary fiber and bone-building manganese. It is very good source of enzyme-catalyzing molybdenum and heart-healthy potassium. Eggplant is also a good source of bone-building vitamin K and magnesium as well as heart-healthy copper, vitamin C, vitamin B6, folate, and niacin. Eggplant also contains phytonutrients such as nasunin and chlorogenic acid.

Phytochemicals

Phytonutrients contained in eggplant include phenolic compounds, such caffeic and chlorogenic acid, and flavonoids, such as *nasunin* (Singh *et al.*, 2009). Nasunins are potent antioxidants, unique for chelating irons (Luthria and Mukhopadhyay, 2006) and can protect lipids in brain cell membranes (Noda *et al.*, 1998). By its ability to remove excess iron, nasunins can reduce free radical formation (Noda *et al.*, 2000). There are two isomers of nasunin, delphinidin 3-[4-(cis-p-coumaroyl)-L-rhamnosyl (1–6) glucopyranoside]-5-glucopyranoside (cis) and delphini-din-3[4- (trans-p-coumaroyl)- L-rhamnosyl-(1–6)glucopyranoside]-5 glucopyranoside (trans) (Das *et al.*, 2011). Two other new isomeric compound were identified by Ma *et al.*, (2010), as 3-O-malonyl-5-O-I-caffeoylquinic acid (isomer 1) and 4-O-I-caffeoyl-5-O-malonylquinic acid (isomer 2) in eggplant fruit. The iron chelation activities of isomers 1 and 2, respectively, were about 3- and 6-fold greater than that of quercetin dihydrate.

Health Benefits of Brinjal or Eggplant fruit

While many fruits and vegetables are important for good health, eggplant appears to play a special role in the treatment and prevention of a number of very serious conditions.

Cancer Prevention: Eggplant has been found to be especially useful in the prevention of colon cancer due to the high amount of fiber found in it. Fiber is a relatively porous nutrient, and as it moves through the digestive tract, it absorbs toxins and chemicals that can lead to the development of colon cancer.

Weight Reduction: Fiber is a relatively "bulky" food, and takes up a lot of room in the stomach. Therefore, by eating eggplant in a salad or appetizer before a meal, dieters are likely to have a greater feeling of satiety, and generally eat fewer calories (thereby achieving a substantial weight loss with time).

Diabetes Management: Egg plant controls the glucose absorption of the body and it also helps reduce hypertension. The glycemic Index of eggplant fruit is low (15), hence good for diabetics.

Cardiac Protection: Eggplant fruit has cardioprotective properties and was found to reduce myocardial ischemia/reperfusion injury. This effect is believed to be brought out through nasunin, the purple anthocyanin in eggplant fruit.

Therapeutic Properties of Brinjal (eggplant)

Das *et al.*, (2011) demonstrated eggplants as containing potent cardioprotective compounds judging by their ability to increase left ventricular function and reduce myocardial infarct size and cardiomyocyte apoptosis in rats.

Significant correlation was found between hepatoprotective activities and total phenolic/flavonoid content (r = 0.6371-0.8842) and antioxidant activities (r = 0.5846-0.9588), of eggfruit plant indicating the contribution of the phenolic antioxidant present in eggplant to its hepatoprotective effect on t-BuOOH-induced toxicity

(*Akanitapichat, et al.*, 2010) . Extracts from purple colour small size eggplant fruit demonstrated higher antioxidant activities than other coloured eggplant fruits (Nisha *et al.*, 2009).

Flavonoids extracted from the fruits of *Solanum melongena* (Brinjal) orally administered at a dose of 1 mg/100 g BW/day showed significant hypolipidemic action in normal and cholesterol fed rats. HMG CoA reductase activity was found to be enhanced, while activities of glucose-6-phosphate dehydrogenase and malate dehydrogenase were significantly reduced. Activities of lipoprotein lipase and plasma LCAT showed significant enhancement. A significant increase in the concentrations of hepatic and faecal bile acids and faecal neutral sterols was also observed indicating a higher rate of degradation of cholesterol (Sudheesh *et al.*, 1997). Powdered fruits of eggplant (*Solanum melongena*), is commercially utilized in Brazil to treat human hyperlipidemia. However, the trials conducted on human subjects in Portugal did not show positive results (Silva *et al.*, 2004).

Studies conducted by Yoshikawa *et al.*, (1996) using Salmonella/microsome assay showed that the Japanese eggplant fruit juice exhibited an antimutagenic activity against 3-amino-1-methyl-5H-pyrido[4,3-b] indole (Trp-P-2) induced mutagenicity.

References

Akanitapichat, P., Phraibung, K., Nuchklang, K., and Prompitakkul, S. 2010. Antioxidant and hepatoprotective activities of five eggplant varieties. Food Chem Toxicol., (2010), 48(10):3017-21.

Das, S., Raychaudhuri, U., Falchi, M., Bertelli, A., Braga, P.C., and Das, D.K. 2011. Cardioprotective properties of raw and cooked eggplant (*Solanum melongena* L). Food Funct., (2011), 2: 395.

Gopalan, C., Sastri, B.V.R. and Balasubramanian, S.C. 1985. Nutritive Value of Indian Foods. National Institue of Nutrition, ICMR, Hyderabad, India.

Luthria, D. L., and Mukhopadhyay, S. 2006. Influence of sample preparation and assay of phenolic acids from eggplant, J. Agric. Food Chem., (2006), 54: 41–47.

Ma, C., Whitaker, B.D., and Kennelly, E. J. 2010. New 5-O-Caffeoylquinic Acid Derivatives in Fruit of the Wild Eggplant Relative Solanum viarum. J Agric Food Chem., (2010), 58(20): 11036-11042.

Nisha, P., Abdul-Nazar, P., and Jayamurthy, P. 2009. A comparative study on antioxidant activities of different varieties of Solanum melongena. Food Chem Toxicol., (2009), 47(10):2640-4

Noda, T Y., Igarashi, K.K., Mori, A., and Packer, L.1998. Antioxidant activity of nasunin, an anthocyanin in eggplant, Res.Commun. Mol. Pathol. Pharmacol., (1998), 102: 175–187.

Noda, T. Y., Igarashi, K. K., Mori, A., and Packer, L. .2000. Antioxidant activity of nasunin, an anthocyanin in eggplant peels, Toxicology, (2000), 148, 119–123.

Silva, G.E., Takahashi, M.H., Eik, F.W., Albino, C.C., Tasim, G.E., Serri, L.A., Assef, A.H., Cortez, D.A., and Bazotte, R. B. 2004. Absence of hypolipidemic effect of Solanum Melongena L. (eggplant) on hyperlipidemic patients. Arq Bras Endocrinol Metabol., (2004), 48(3):368-73.

Singh, A. P., Luthria, D., Wilson, T., Vorsa, N., Singh, V., Banuelos, G.S., and Pasakdee, S. 2009. Polyphenols content and antioxidant capacity of eggplant pulp, Food Chem., (2009), 114: 955–961.

Sudheesh, S., Presannakumar, G., Vijayakumar, S., and Vijayalakshmi, N.R. 1997. Hypolipidemic effect of flavonoids from Solanum melongena. Plant Foods Hum Nutr., (1997), 51(4):321-30.

Yoshikawa, K., Inagaki, K., Terashita, T., Shishiyama, J., Kuo, S., and Shankel, D.M. 1996. Antimutagenic activity of extracts from Japanese eggplant. Mutat Res., (1996), 371(1-2):65-71.

4.3. Chillies & Capsicum (*Capsicum annuum*, L., or *Capsicum frutescens*)

The chillies (green or dried ripe) belongs to family Solanaceae. Botanical name of chilli is Capsicum *annum L.*, and sometiens *Capsicum frutescens*. It is used for its pungent and spicy taste besides the appealing colour it adds to the food. Paprikas or bell pepper (capsicum) are mainly the European cultivars which are large sized mild fruits belonging to this species. On account of discoveries made in burial tombs in Peru, the origin of capsicum is now agreed to be of the new world. The centre of diversity of the common cultivated pepper (Capsicum *annum*) is probably Mexico and the secondary centre is Guatemala. The annual plant types belong to *Capsicum annuum*, the perennial plant types belong to *Capsicum frutescens*. In India, black pepper (*Piper nigrum*), was used in place of chillies before it's introduction. Today its use is so wide that no curry preparation goes with chilli (green or red dried). The pungent forms are used as chilli, whole dry chilli, chilli powder, chilli paste, chilli sauce, chilli oleoresin or as mixed curry powder. Dried fruits are extensively used as spice. India contributes about 36% to the total world production. In India, chillies are grown in almost all the states through out the country. Andhra Pradesh is the largest producer of chilli in India and contributes about 26% to the total area under chilli, followed by Maharashtra (15%), Karnataka (11%), Orissa (11%), Madhya Pradesh (7%) and other states contributing nearly 22% to the total area under chilli.

Varieties: Jwala, X-235, G-1, G-2, G-3, G-4, G-5, LCA-205, 206, 235, Karakulu, Sannalu, Dippayerupu, Punasa, Maduru, Pottibudaga, Hybrid, Bharat, Aparna, Pottikayalu, Cullakayalu, Barak, Mota and Chapta are widely cultivated varieties in Andhra Pradesh. Jwala, Bayadgi, G-1, G-2, G-3, G-4, G-5 and Pusa Jwala are cultivated in Karnataka. K-1, K-2, CO-1, CO-2, CO-3, PMK-1, PMK-2 and Borma are popular in Tamil Nadu and Kerala. Pathori, Bugayati, Dhobri, Black seed, Chaski, Bhiwapuri, Pusa Jwala, Sona-21, Jawahar, Sadabahar and Agni are widely grown in Central India. Rori, Moti Mirchi, Chittee, NP-46-A, Pusa Jwala, Pusa Summer, Solan Yellow, Hot Portugal, Pachad Yellow, Sweet Banana, Hungarian Wax, Punjab Lal, CH-1, Sanauri, NP-46, Jwala Pant C-1, Desh, Pahadi, Kalyanpur, Chaman and Chanchal are the widely grown varieties of northern region. Arka Meghana, Arka Kyathi, Arka Harita, Arka Suphal, Arka Swetha and Arka Lohit are some varieties released from ICAR-IIHR, Bengaluru and spreading in Southern and Eastern India.

Uses

Chillies are used throughout the tropics as a major ingredient of curry powders. Extracts of chillies are used in the production of ginger beer and other beverages. Cayenne pepper is incorpotated in poultry feeds while *C. frutesens* is used for its carminative medicinal properties. Green chillies are rich in 'rutin' which has immense medicinal value. The substances that give chilli their intensity when ingested or applied topically are *capsaicin* (8-methyl-*N*-vanillyl-6-nonenamide) and several related chemicals, collectively called *capsaicinoids* (Kosuge *et al.*, 1961). Capsaicin is the primary ingredient in the pepper spray used as an irritant weapon.

Pure capsaicin is a hydrophobic, colorless, odorless, and crystalline-to-waxy solid at room temperature. As far as its medicinal use is concerned capsaicin is reported to be a safe and effective topical analgesic agent in the management of arthritis pain, herpes zoster-related pain, diabetic neuropathy, postmastectomy pain, and headaches. It is rich in vitamin C, phosphorus, calcium and magnesium.

Nutritional Components of Hot Chilli and Bell Pepper

Hot chilli [#][*]		Sweet Bell Pepper[#]
Component	**Quantity per 100g**	**Quantity per 100g**
*Moisture	88.23±1.40 g	93.89
*Carbohydrates	2.71±0.06 g	4.64
*Protein	0.91±0.02 g	0.86
*Fat	1.18±0.06 g	0.17
*Fibre	6.25±0.38 g	1.7
*Ash	0.75±0.06 g	0.43
*Iron	0.79±0.06 mg	0.34 mg
*Zinc	0.18±0.02 mg	0.13 mg
*Phosphorus	34.66±0.98 mg	20 mg
*Copper	0.16±0.01 mg	0.07 mg
*Manganese	0.21±0.02 mg	0.12
*Calcium	16.04±0.78 mg	10 mg
*Magnesium	20.03±0.86 mg	10 mg
[#]Vitamin C	80.6 mg	80.6 mg
[#]Vitamin A	18 mg	18 µg
[#]Niacin	0.48 mg	0.48 mg
[#]Vitamin B6	0.224 mg	0.03

*Ananth et al., (2014) ; [#] Olatunji and Afolayan (2018)

Phytochemicals

Fresh green chillies are excellent source of vitamin C while red chillies are rich in carotene, cryptoxanthin and lycopene. It has good amount of fibre and good source of most B vitamins, and vitamin B_6 in particular. They are very high in potassium, magnesium, and iron. Their high vitamin C content can also substantially increase the uptake of non-heme iron from other ingredients in a meal, such as beans and grains. Capsaicin is a remarkable health-promoting substance present in chillies.

Two steroidal saponins were isolated and purified from cayenne pepper (*Capsicum frutescens*) by De Lucca et al., (2006) and they were found to have fungicidal properties when tested against *Aspergillus flavus, A. niger, A. parasiticus, A. fumigatus, Fusarium oxysporum, F. moniliforme, and F. graminearum.*

The most abundant constituents present in the volatile fractions of three varieties of chilli peppers were esters and alcohols in the malagueta chilli pepper (*C. frutescens*), monoterpenes and aldehydes in the dedo-de-moc‚a chilli pepper (*C. baccatum* var. pendulum), and esters and sesquiterpenes in the murupi chilli pepper (*C. chinense*) (Junior *et al.*, 2012). But most of the volatile compounds identified by Gurnani *et al.*, (2016) were hydrocarbons, fatty acids, fatty esters, and some novel constituents, such as *β*-diketones. Many of these identified compounds have already been reported to be pharmacologically active.

Therapeutic Properties of Chillies and Bell Peppers

Natural Pain Relief: Topical capsaicin is now a recognized treatment option for osteoarthritis pain. Several review studies of pain management for diabetic neuropathy have listed the benefits of topical capsaicin to alleviate disabling pain associated with this condition.

Cardiovascular Benefits: Red chilli have been shown to reduce blood cholesterol, triglyceride levels, and platelet aggregation, while increasing the body's ability to dissolve fibrin, a substance integral to the formation of blood clots. Eating freshly chopped chilli was found to increase the resistance of blood fats, such as cholesterol and triglycerides, to oxidation (free radical injury).

Clear Congestion: Capsaicin not only reduces pain, but its peppery heat also stimulates secretions that help clear mucus from stuffed up nose or congested lungs.

Improves Immunity: Chillies are rich in carotenoids (which converts into vitamin A) and vitamin C. Often called the anti-infection vitamin, vitamin A is essential for healthy mucous membranes, which line the nasal passages, lungs, intestinal tract and urinary tract and serve as the body's first line of defense against invading pathogens.

Prevent Stomach Ulcers: Chillies have a bad–and mistaken–reputation for contributing to stomach ulcers. Not only that they do not cause ulcers, but they can help prevent them by killing bacteria that has been ingested, while stimulating the cells lining the stomach to secrete protective buffering juices.

Anti-inflammatory: Pharmacological and physiological studies demonstrated that Capsaicin, which contains a vanillyl moiety, produces its sensory effects by activating a Ca2 +-permeable ion channel on sensory neurons. Capsaicin is a known activator of vanilloid receptor 1. Capsaicin-induced stimulation of prostaglandin biosynthesis has been shown using bull seminal vesicles and rheumatoid arthritis synoviocytes. Capsaicin inhibits protein synthesis in Vero kidney cells and human neuroblastoma SHSY-5Y cells *in vitro*. Capsaicin is a potent inhibitor of substance P, a neuropeptide associated with inflammatory processes. Capsaicin is being studied as an effective treatment for sensory nerve fiber disorders, including pain associated with arthritis, psoriasis, and diabetic neuropathy. When animals injected with a substance that causes inflammatory arthritis were fed a diet that contained capsaicin, they had delayed onset of arthritis, and also significantly reduced paw inflammation.

Anti-Microbial: Capsaicin inhibited growth of *E. coli, Pseudomonas solanacearum,* and *Bacillus subtilis* bacterial cultures, but not *Saccharomyces cerevisiae* (Anonymous, 2007). Rebeiro *et al.,* (2007) isolated and characterized peptides present in chilli seeds (*Capsicum annuum* L.) and evaluated their toxic activities against some yeast species, found that it can prevent the growth and multiplication of yeasts. In the determination of the *in vitro* antimicrobial activity, seed extracts prevented the growth of most of the tested pathogens by forming significant inhibition zones. The inhibitory activity was especially remarkable (inhibition zone 13 mm) against *Pesudomaonas aeruginosa, Klebsilla pneumonae, Staphylococcus aureus* and *Candida albicans.* During the evaluation of the *in vitro* antioxidant activity *via* DPPH assay, *n*-hexane and chloroform extracts showed 26.9% and 30.9% free radical scavenging abilities, respectively, at the concentration of 1 mg/mL. Considering these results, *C. frutescens* seeds can be used as a source of novel antimicrobial and antioxidant compounds (Gurnani *et al.,* 2012).

Anti-Diabetic property: Insulin required to lower blood sugar after a meal is reduced if the meal contains chilli. In overweight people, not only do chilli-containing meals significantly lower the amount of insulin required to lower blood sugar levels after a meal, but chilli-containing meals also result in a lower ratio of C-peptide/ insulin, an indication that the rate at which the liver is clearing insulin has increased.

For the first time the phenols, flavonoids, carotenoids, capsaicin and dihydrocapsaicin content of *Capsicum annuum* var. *acuminatum* and their antioxidant and hypoglycemic properties was evaluated by Tundis *et al.,* (2011). They found the highest radical scavenging activity in ethanol extracts of *Capsicum annuum* var. *acuminatum.* The lipophilic fraction of both *C. annuum* var. acuminatum and *C. annuum* var. cerasiferum exhibited an interesting and selective inhibitory activity against α-amylase (Tundis *et al.,* 2011; Loizzo *et al.,* 2008). Oboh *et al.,* (2011) also found the ability of the chilli extracts to inhibit key enzymes linked with type 2 diabetes (α-amylase and α-glucosidase). Purified capsaicin caused a decrease in blood glucose levels during glucose tolerance tests in dogs. There was a concomitant elevation in plasma insulin levels indicating its role in treating type 2 diabetes (Tolan *et al.,* 2004).

Antioxidant property: The studies conducted by Kim *et al.,* (2011) indicated that the amounts of capsanthin and L-ascorbic acid in red paprika fruits correlated well with its antioxidant activity. Paprika leaves, which has various phytochemicals such as lutein, chlorophyll, and γ-tocopherol, might be used in nutraceuticals and pharmaceuticals for improving human health. Oboh and Ogunruku (2009) showed that dietary hot short pepper (*Capsicum frutescens* L. var. abbreviatum) could prevent cyclophosphamide-induced oxidative stress in brain; and opined that the protective effect of the pepper could be attributed to their antioxidant properties.

Anti-Cancerous: Capsaicin stops the spread of prostate cancer cells through a variety of mechanisms. Capsaicin triggers suicide in both primary types of prostate cancer cell lines and also lessens the expression of prostate-specific antigen (PSA) directly by inhibiting PSA transcription, causing PSA levels to plummet. Intratumoral administration of capsaicin into a pre-existing tumor resulted in retarded progression of the injected tumor regardless of whether the tumor is at its early or late stage (Beltran *et al.*, 2007). Furthermore, it led to significant inhibition of growth of other, uninjected tumors in the same animal. Capsaicin-elicited immunity is shown to be T cell-mediated and tumor-specific. These results reflect the immunological potency of a neurological ligand in modulating immune response against an established tumor

References

Ananthan, R., Subhash, K., and Longvah, T. 2014. Assessment of nutrient composition and capsaicinoid content of some red chillies. International Proceedings of Chemical, Biological and Environmental Engineering, (2014)72(1): 1-4.

Anonymous. 2007. Final report on the safety assessment of *Capsicum annuum* extract, *Capsicum annuum* fruit extract, *Capsicum annuum* resin, *Capsicum annuum* fruit powder, *Capsicum frutescens* fruit, *Capsicum frutescens* fruit extract, *Capsicum frutescens* resin, and capsaicin. Int J Toxicol. 2007;26 Suppl 1:3-106.

Anonymous. 2009. Post Harvest Profile of chilli. Directorate of Marketing and Inspection, Nagpur. Ministry of Agriculture, DAC, Government of India.

Beltran, J., Ghosh, A.K., and Basu, S. 2007. Immunotherapy of tumors with neuroimmune ligand capsaicin. J Immunol., (2007), 178(5):3260-4.

De Lucca, A.J., Boue, S., Palmgren, M.S., Maskos, K., and Cleveland, T.E. 2006. Fungicidal properties of two saponins from *Capsicum frutescens* and the relationship of structure and fungicidal activity. Can J Microbiol., (2006), 52(4):336-42.

Gurnani, N., Gupta, M., Mehta, D., and Mehta, B.K. 2016. Chemical composition, total phenolic and flavonoid contents, and in-vitro antimicrobial and antioxidant activities of crude extracts from red chilli seeds (Capsicum frutescens L.). Journal of Taibbah University for Science, (2016), 10:462-470.

Junior, S.B., Tavares, A.M., Filho, J.T., Zuni, C.A., and Godoy, H.T. 2012. Analysis of the volatile compounds of Brazilian chilli peppers (Capsicum spp.) at two stages of maturity by solid phase micro-extraction and gas chromatography-mass spectrometry. Food Res. Int., (2012), 48:98-107.

Kim, J.S., Ahn, J., Lee, S.J., Moon, B., Ha, T.Y., and Kim, S. 2011. Phytochemicals and antioxidant activity of fruits and leaves of paprika (*Capsicum Annuum* L., var. special) cultivated in Korea. J Food Sci., (2011), 76(2):C193-8. Doi: 10.1111/j.1750-3841.2010.01891. x.

Kosuge, S., Inagaki, Y., and Okumura, H. 1961. Studies on the pungent principles of red pepper. Part VIII. On the chemical constitutions of the pungent principles. Nippon Nogei Kagaku Kaishi (J. Agric. Chem. Soc.), 35, 923–927; (en) Chem. Abstr. 1964, 60, 9827g.

Loizzo, M.R., Tundis, R., Menichini, F., Statti, G.A., and Menichini, F. 2008. Influence of ripening stage on health benefits properties of *Capsicum annuum* var. *acuminatum* L.: *in vitro* studies. J Med Food., (2008), 11(1):184-9.

Oboh, G., Ademiluyi, A.O., and Faloye, Y.M. 2011. Effect of combination on the antioxidant and inhibitory properties of tropical pepper varieties against α-amylase and α-glucosidase activities *in vitro*. J Med Food., (2011), 14(10):1152-8.

Oboh, G., and Ogunruku, O.O. 2009. Cyclophosphamide-induced oxidative stress in brain: protective effect of hot short pepper (*Capsicum frutescens* L. var. *abbreviatum*). Exp Toxicol Pathol., (2010), 62(3):227-33.

Olatunji, T.L., and Afolayan, A. 2018. The suitability of chilli pepper (Capsicum annum L.) for alleviating human micronutrient dietary deficiencies: A review. Food Sci Nutr., (2018), 6: 2239-2251.

Ribeiro, S.F., Carvalho, A.O., Da Cunha, M., Rodrigues, R., Cruz, L.P., Melo, V.M., Vasconcelos, I.M., Melo, E.J., Gomes, V.M. 2007.Isolation and characterization of novel peptides from chilli pepper seeds: antimicrobial activities against pathogenic yeasts. Toxicon., (2007), 50(5):600-11.

Tolan, I., Ragoobirsingh, D., and Morrison, E.Y. 2004. Isolation and purification of the hypoglycaemic principle present in *Capsicum frutescens*. Phytother Res., (2004), 18(1): 95-6.

Tundis, R., Loizzo, M.R., Menichini, F., Bonesi, M., Conforti, F., Statti, G., De Luca, D., de Cindio, B., and Menichini, F. 2011. Comparative study on the chemical composition, antioxidant properties and hypoglycaemic activities of two *Capsicum annuum* L. cultivars (Acuminatum small and Cerasiferum). Plant Foods Hum Nutr., (2011), 66(3):261-9.

5. Root Crops

Carrot, radish, turnip and beet root are major root crops grown in India. In addition, root crops like rutabaga, parsnip, parsley, chervil and celeriac are grown in a limited scale in different parts of the world. There are several other roots and modified underground plant parts that grow in forests and consumed by tribes, but they are not commercially cultivated. Carrot, radish, turnip and beet root are generally consumed as cooked vegetables or fresh salads or as pickled vegetables.

5.1. Carrot

Carrots belong to the *Umbelliferae* family, named after the umbrella-like flower clusters that plants in this family produce. As such, carrots are related to parsnips, fennel, parsley, anise, caraway, cumin and dill. Carrot, *Daucus carota* L. is believed have been primarily originated in Afghanistan, South western Asia and Mediterranean region. *Daucus carota* ssp. *Carota* is the most common wild form in Europe and South West Asia, and the present day cultivated carrots have most probably originated from this sub-species.

Haryana, Andhra Pradesh, Punjab and Tamil Nadu, Karnataka, Uttar Pradesh and Assam are the major producers of carrot in India.

Cultivars/Varieties: Varieties with long, orange coloured and smooth roots are preferred in India. Both indigenous and exotic varieties having different root length, shape and colour are grown. Temperate or European type varieties are Nantes, Half long, Early Nantes, Chantenay, Chaman, Pusa Yamadagni and Ooty-1, which require low temperature of 4-8°C for flowering. Tropical or Asiatic type varieties do not require low temperature for flowering and can be grown in plains also. The varieties belonging to this group are Pusa Kesar, Pusa Meghali and Hisar Gairic. Some varieties like Gold King, Indian Kuroda and Super Kuroda are also marketed by private seed companies.

Uses

Carrots are used as a vegetable for soups, stews, curries and pies; grated carrots are used as salad, tender roots as pickles. Gajar halwa is a delicious dish popular in northern part of India. Carrot juice is a rich source of carotene and is sometimes used for colouring butter and other foods. Carrot is valued as a nutritive food mainly due to its high content of alph and beta-carotene. The nutritive value of carrot is enlisted below.

Besides its use as a vegetable it is also used for its medicinal value in folk medicine. An infusion of carrot has been used as disinfesting agent for threadworms. The essential oil extracted from carrot roots is said to have antibacterial properties (Schuphan and Weiller, 1967). Carrot consumption increases the quantity of urine and helps in elimination of uric acid. Addition of large amount of carrot to the diet has a favourable effect on the nitrogen balance. They are reported to be useful in diseases of the kidney and in dropsy (Chopra, 1933; Kirtikar and Basu, 1935).

Nutritional Components of Carrot Roots		Nutritional Components in Carrto Greens	
Constituent	Quantity g/100g	Component	Quantity g/100g
Carbohydrates	10.6 g	Carbohydrates	8.3 g
Protein	0.9 g	Protein	5.1 g
Fat	0.2 g	Fat	0.5 g
Moisture	86 g	Moisture	83.9 g
Fibre	1.20 g	Minerals	2.8
Minerals	1.1 g	Phosphorus	110 mg
Iron	2.2 mg	Iron	8.8 mg
Carotene	1.89 mg	Calcium	340 mg
Thiamine	0.04 mg		
Riboflavin	0.02 mg		
Niacin	0.5 mg		
Vitamin C	3.0 mg		
Folic acid	15 ug		
Calcium	80 mg		
Phosphorus	30 mg		
Energy	48 Kcal		

Phytochemicals

All varieties of carrots contain valuable amounts of antioxidant nutrients. Included here are traditional antioxidants like vitamin C, as well as phytonutrient antioxidants like beta-carotene. The phytochemical in carrot includes

- Carotenoids – alpha-carotene, beta-carotene, lutein
- Hydroxycinnamic acids – caffeic acid, coumaric acid, ferulic acid
- Anthocyanindins – cyanidins, malvidins

Red and purple carrots, for example, are best known for the rich anthocyanin content. In yellow carrots, 50% of the total carotenoids come from lutein. Polyacetylenes are another category of phytonutrients which is believed to confer protection against cardiovascular disease. Polyacetylenes are unique phytonutrients made from metabolism of fatty acids crepenynic acid, stearolic acid and tariric acid. They are common in the members of *Apiaceae/Umbelliferae* family of plants (which includes carrot). The two best-researched polyacetylenes in carrot are falcarinol and falcarindiol.

Health Benefits of Carrots

Cardiovascular system needs constant protection from antioxidant damage. Antioxidant nutrients in carrots are believed to explain many of the cardioprotective benefits provided by these root vegetables. The many different kinds of carrot antioxidants are most likely to work together and provide us with cardiovascular benefits.

- Carrots are rich in β-carotene, that is the precursor for vitamin A. Small scale human studies show clear benefits of carrot intake for eye health. Studies at the Jules Stein Institute at the University of California, Los Angeles showed that women who consumed carrots at least twice per week – in comparison to women who consume carrots less than once per week – have significantly lower rates of glaucoma (damage to the optic nerve often associated with excessive pressure inside the eye).

- Lab studies have shown the ability of carrot extracts to inhibit the growth of colon cancer cells. The polyacetylenes found in carrot (especially falcarinol) have been specifically linked to this inhibitory effect. In studies of carrot juice intake, it was found that those who consumed 1.5 cups of carrot juice per day showed good colon cell health compared to those who did not.

- An infusion of carrot seeds (1 teaspoon per cup of boiling water) is believed to be diuretic, to stimulate the appetite, reduce colic, aid fluid retention and help alleviate menstrual cramps.

- The dried flowers are also used as a tea as a remedy for dropsy.

- The seeds if carrots boiled in water (seed tea) or wine when taken has served as contraceptive.

- When applied with honey, the leaves of carrot were said to cleanse running sores or ulcers.

- Chewing a carrot immediately after food kills all the harmful germs in the mouth. It cleans the teeth, removes the food particles lodged in the crevices and prevents bleeding of the gums and tooth decay.

- Carrot soup is supposed to relieve diarrhoea and help with tonsilitus.

- Carrots as soup or juice is said to improve memory and relieve nervous tension.

The alternative medicine believers consider the carrot (the whole plant or its seeds) to have the following properties:

- Anthelmintic *(destroying or expelling worms)*.
- Carminative *(expelling flatulence)*.
- Contraceptive.
- Deobstruent.
- Diuretic *(promoting the discharge of urine)*.
- Emmenagogue *(producing oils which stimulate the flow of menstrual blood)*.
- Galactogogue *(promoting the secretion of milk)*.
- Ophthalmic *(pertaining to the eye)*.
- Stimulant.
- Oedema *(water retention)*.

Therapeutic Properties of Carrots

Because of their antioxidant properties, carotenoids may have beneficial effects in preventing cancer and cardiovascular disease. In a study conducted by Vasudevan and Parle (2006), the ethanolic extract of *Daucus carota* seeds (DCE) was administered orally in three doses (100, 200, 400 mg/kg) for seven successive days to different groups of young and aged mice. The extent of memory improvement evoked by DCE was 23% at the dose of 200 mg/kg and 35% at the dose of 400 mg/kg in young mice using elevated plus maze. Similarly, significant improvements in memory scores were observed using passive avoidance apparatus and aged mice. Furthermore, DCE reversed the amnesia induced by scopolamine (0.4 mg/kg, i.p.) and diazepam (1 mg/kg, i.p.). *Daucus carota* extract (200, 400 mg/kg, *p.o.*) reduced significantly the brain acetylcholinesterase activity and cholesterol levels in young and aged mice.

It has been observed that elderly patients suffering from Alzheimer's disease showed reduction in symptoms of disease upon chronic use of anti-inflammatory drugs (Rao *et al.*, 2002; Stephan *et al.*, 2003). Epidemiological studies have almost confirmed that non- steroidal anti-inflammatory drugs reduced the incidence of AD (Rao *et al.*, 2002; Stephen *et al.*, 2003 & Breitner, 1996). Compounds such as Geraniol, 2,4,5-trimethoxy benzaldehyde (TMB), oleic acid and transasarone isolated from *Daucus carota* seeds have been shown to possess anti-inflammatory action in rodents. (Momin *et al.*, 2003). The quaternary base chlorides separated from the seeds of *Daucus carota* were rich in choline content and exhibited procholinergic activity (Gambhir *et al.*, 1966 a & b). Thus, it is possible that enhanced cholinergic transmission resulting from increased acetylcholine synthesis in brain due to abundant availability of choline and reduction of brain cholinesterase activity in young and aged mice may explain the memory improving effect exhibited by DCE.

Among carotenoids, TEAC (Trolox equivalent antioxidant capacity) values are highest for lycopene, β-carotene and lutein (Miller *et al.*, 1996). Because antioxidant mechanisms are likely involved in the pathogenesis of cardiovascular diseases and cancer (Gaziano and Hennekens 1993, Ziegler 1991), it is assumed that the health benefits associated with the consumption of carotenoid rich vegetables are due to atleast in part to antioxidant properties of the carotenoids. This hypothesis is supported by studies that showed an increase in plasma antioxidant capacity in humans (Cao *et al.*, 1998a and 1998b, Miller *et al.*, 1998) and protection against lipid peroxidation as measured by thiobarbituric acid reactive substances (TBARS) and breath pentane (Miller *et al.*, 1998) upon increased consumption of fruit and vegetables.

Carrots have been suggested as a potential treatment for leukaemia in traditional medicine and have previously been studied in other contexts as potential sources of anticancer agents. Zaini *et al.*, (2012) investigated the effects of five fractions from carrot juice extract (CJE) on human lymphoid leukaemia cell lines, together with five purified bioactive compounds found in *Daucus carota* L, including: three polyacetylenes (falcarinol, falcarindiol and falcarindiol-3-acetate) and two carotenoids (beta-carotene and lutein). Their effects on induction of apoptosis was studied using Annexin V/PI and Caspase 3 activity assays and inhibition of cellular proliferation using Cell Titer Glo assay and cell cycle analysis were investigated. Treatment of all three lymphoid leukaemia cell lines with the fraction from carrot extracts which contained polyacetylenes and carotenoids was significantly more cytotoxic than the 4 other fractions. Treatments with purified polyacetylenes also induced apoptosis in a dose and time responsive manner. Moreover, falcarinol and falcarindiol-3-acetate isolated from *Daucus carota* L were more cytotoxic than falcarindiol. In contrast, the carotenoids showed no significant effect on either apoptosis or cell proliferation in any of the cells investigated. This suggests that polyacetylenes rather than beta-carotene or lutein are the bioactive components that could be useful in the development of new leukemic therapies. Here, for the first time, the cytotoxic effects of polyacetylenes have been shown to be exerted via induction of apoptosis and arrest of cell cycle.

References

Breitner, J.C.S. 1996. The role of anti-inflammatory drugs in the prevention and treatment of Alzheimers disease [Review]. Ann. Rev. Med., 47: 401-11.

Cao, G. H., Russell, R. M., Lischner, N., and Prior, R. L. 1998b. Serum antioxidant capacity is increased by consumption of strawberries, spinach, red wine or vitamin C in elderly women. J. Nutr., 128: 2383-2390.

Cao, G.H., Booth, S. L., Sadowski, J. A., and Prior, R. L. 1998a. Increases in human plasma antioxidant capacity after consumption of controlled diets high in fruit and vegetables. Am. J. Clin. Nutr., 68: 1081–1087.

Chopra, R.N. 1933. Indigenous Drugs of India. The Art Press, Calcutta.

Gambhir, S,S., Sanyal, A.K., Sen, S.P., and Das, P.K. 1966. Studies on *Daucus carota*, Linn. I. Pharmacological studies with the water-soluble fraction of the alcoholic extract of the seeds: a preliminary report. Indian J Med Res., (1966), 54(2):178-87.

Gambhir, S. S., Sanyal, A. K., Sen, S. P., and Das, P.K. 1966. Studies on *Daucus carota* Linn. Part II. Cholinergic activity of the quaternary base isolated from water-soluble fraction of alcoholic extract of seeds. Indian J. Med. Res., 54: 1053-1056.

Gaziano, J. M. and Hennekens, C. H. 1993. The role of b-carotene in the prevention of cardiovascular disease. Ann. N.Y. Acad. Sci., 691: 148-155.

Kirtikar, K.R. and Basu, B.D.1995. Indian Medicinal Plants (Iolit Mohan Basu), Allahabad.

Miller, E. R., Appel, L. J., and Risby, T. H. 1998. Effect of dietary patterns on measures of lipid peroxidation—results from a randomized clinical trial. Circulation, 98: 2390–2395.

Miller, N. J., Sampson, J., Candeias, L. P., Bramley, P. M. and Rice-Evans, C. A. 1996. Antioxidant activities of carotenes and xanthophylls. FEBS Lett., 384: 240-242.

Momin, R.A., De Witt, D.L., Nair, M.G. 2003. Inhibition of cyclooxygenase (COX) enzymes by compounds from *Daucus carota* L. Seeds. Phytother Res., (2003), 17(8):976-9.

Rao, S. K., Andrade, C., Reddy, K., Madappa, K. N., Thyagarajan, S., and Chandra, S. 2002. Memory protective effect of indomethacin against electroconvulsive shock-induced retrograde amnesia in rats, Biological Psychiatry, 51(9): 770-773. Doi:10.1016/S0006-3223(01)01219-7

Sadhu, M. K and Sarkar, K. 2003.Carrot. In: Vegetable Crops Vol 2. (Eds. T.K. Bose., Kabir.J., Maity, T.K., Parthasarathy, V.A. and Som, M.G)., Naya Udyog Publishers, Kolkatta. pp.1-48.

Schuphan, W and Weiller, H. 1967. Qual. Plan. Mater.Veg., 15:81-101.

Stephan A., Laroche S., Davis S., 2003. Eur. J. Neurosci., 17: 1921—1927 (2003).

Vasudevan, M. and Parle, M. 2006. Pharmacological evidence for the potential of *Daucus carota* in the management of cognitive dysfunctions. Biol. Pharm. Bull. **29**(6) :1154—1161.

Zaini, R.G., Brandt, K., Clench, M.R., Le Maitre, C.L. 2012.Effects of bioactive compounds from carrots (*Daucus carota* L.), polyacetylenes, beta-carotene and lutein on human lymphoid leukaemia cells. Anticancer Agents Med Chem., (2012),12(6): 640-652.

Ziegler, R. 1991. Vegetables, fruits, and carotenoids and the risk of cancer. Am. J. Clin. Nutr., 53: 251S–259S.

5.2. Beet Root (*Beta vulgaris* L.)

The beet, is one of the many cultivated varieties of *Beta vulgaris*. Beet root, sugar beet and palak belong to the same species *Beta vulgaris* and are cross compatible. Beet root is has evolved from *Beta vulgaris* L. ssp. *maritima* by hybridization with *Beta patula*. The origin of this crop is believed to be in Europe.

The garden beet or table beet is a popular root vegetable which is eaten boiled or as a salad. The tops or greens are used as animal feed. Beet greens are rich in iron and vitamins.

Cultivar/Varieties: The improved varieties of beet root popularly cultivated in India are Detroit Dark Red, Crimson Globe, Early Wonder, Ooty-1, Crosy Egyptian, Madhur, Ruby Queen and Ruby Red.

Nutritional Components of Beet root

Constituent	Quantity (per 100 g)	Constituent	Quantity (per 100 g)
Moisture	87.7 g	Riboflavain	0.09 mg
Carbohydrates	8.8 g	Vitamin C	88 mg
Proteins	1.7 g	Calcium	200 mg
Fats	0.1 g	Phosphorus	55 mg
Minerals	0.8 g	Iron	1.0 mg
Thiamine	0.04 mg	Potassium	43 mg

Phytochemicals

Beetroot is a rich source of potent antioxidants and nutrients, including magnesium, sodium, potassium and vitamin C, and betaine, which is important for cardiovascular health. It functions by acting with other nutrients to reduce the concentration of homocysteine, a homologue of amino acid cysteine, which can be harmful to blood vessels and thus contribute to the development of heart disease, stroke, and peripheral vascular disease. Betaine functions in conjunction with S-adenosylmethionine, folic acid, and vitamins B_6 and B_{12} to carry out this function. Yellow varieties are rich in ß-xanthin pigment. Betacyanin in beetroot is a powerful antioxidant which fights against various cancers. Beetroot is high in folate which supports red blood cell growth and helps to prevent anemia. Beetroot is also high in potassium which helps the nerves and muscles to function properly. Potassium also maintains the bodies' acid balance and can help lower the risk of high blood pressure. Beet greens (tops) are an excellent source of carotenoids, flavonoid antioxidants, and vitamin A. Consumption of natural vegetables rich in flavonoids helps to protect from lung and oral cavity cancers. Betalains are water-soluble plant pigments present in beetroot that are widely used as food colorants, and have a wide range of desirable biological activities, including antioxidant, anti-inflammatory, hepatoprotective, anti-cancer properties. Betaine is a modified amino acid consisting of glycine with three methyl groups that serves as a methyl donor in several metabolic pathways and is used to treat the rare genetic causes of homocystinuria. The high antioxidant activity of the hairy root extracts (beet root) was associated with increased concentrations (more than 20-fold) of total phenolic concomitant compounds, which may have synergistic effects with betalains. The presence of 4-hydroxybenzoic acid, caffeic acid, catechin hydrate, and epicatechin were detected in hairy root extracts of beet root. Rutin was only present at high concentration (1.096 mg/g dry extract) in betalain extracts from the hairy root cultures, whereas chlorogenic acid was only detected at measurable concentrations in extracts from intact plants (Georgiev *et al.*, 2010).

Health Benefits of Beet Root

Beets have long been known for its amazing health benefits for almost every part of the body.

Acidosis: Beetroot is considered as alkaline in nature, hence is effective in combating acidosis.

Anemia: Beetroot has around 1.0 mg/100g iron content which regenerates and reactivates the red blood cells and supplies fresh oxygen to the body. The copper content in beets help make the iron more available to the body.

Blood pressure: Its potassium content and other healing and medicinal values effectively normalizes blood pressure.

Cancer: Betaine, an amino acid in beetroot, has significant anti-cancer properties. Studies show that beets juice inhibits formation of cancer-causing compounds and is protective against colon or stomach cancer.

Constipation: The cellulose content helps to ease bowel movements. Drinking beets juice regularly will help relieve chronic constipation.

Detoxification: The choline from this wonderful juice detoxifies not only the liver, but also the entire system of excessive alcohol abuse.

Gastric ulcer: Mix honey with your beets juice and drink two or three times a week on an empty stomach (more frequently if your body is familiar with beets juice). It helps speed up the healing process of gastric ulcer.

Gall bladder and kidney ailments: Beetroot juice coupled with carrot juice is a superb cleansing agent and exceptionally good for curing ailments relating to these two organs.

Liver or bile: The cleansing virtues in beets juice is very healing for liver toxicity or bile ailments, like jaundice, hepatitis, food poisoning, diarrhea or vomiting. A squeeze of lime with beets juice heightens the efficacy in treating these ailments.

Varicose veins: Regular consumption of beetroot juice is believed to help varicose vein problem.

Therapeutic properties of Beetroot

Diets rich in fruits and vegetables reduce blood pressure (BP) and the risk of adverse cardiovascular events. However, the mechanisms of this effect were not elucidated till recently. Certain vegetables like beetroot possess a high nitrate content, and it is hypothesized that this might represent a source of vasoprotective nitric oxide via bioactivation. Peripheral arterial disease (PAD) results in a failure to adequately supply blood and oxygen (O_2) to working tissues and presents as claudication pain during walking. Nitric oxide (NO) bioavailability is essential for vascular health and function. Kenjale et al., (2011) hypothesized that dietary supplementation of nitrate in the form of beetroot (BR) juice would increase plasma NO_2 concentration and increase exercise tolerance. Their experimentation proved that NO_2(-)-related NO signaling increases peripheral tissue oxygenation in areas of hypoxia and increases exercise tolerance in PAD. The studies by Rokkedal-Lausch et al., (2019) provides new evidence that chronic high-dose NO_3^- supplementation through beetroot juice improves cycling performance of well-trained cyclists in both normoxia and hypoxia. The study suggests that the effects of NO_3^- are augmented during conditions of reduced oxygen availability (e.g., hypoxia), thereby increasing the probability of performance improvements for well-trained athletes in hypoxia vs. normoxia.

Ingestion of dietary (inorganic) nitrate elevates circulating and tissue levels of nitrite via bioconversion in the entero-salivary circulation. In addition, nitrite is a potent vasodilator in humans, an effect thought to underlie the blood pressure-lowering effects of dietary nitrate (in the form of beetroot juice) ingestion. The findings of Kapil et al., (2010) demonstrated dose-dependent decreases in blood pressure and vasoprotection after inorganic nitrate ingestion in the form of either supplementation or by dietary elevation using beetroot juice.

Previous cancer chemoprevention studies from the Department of Pharmaceutical Sciences, Howard University, Washington has demonstrated that the extract of red beetroot (Beta vulgaris L.), can be effective in suppressing the development of multi-organ tumors in experimental animals. The scientists (Kapadia et al., 2011) compared the cytotoxic effect of the red beetroot extract with anticancer drug, doxorubicin (adriamycin) in the androgen-independent human prostate cancer cells (PC-3) and in the well-established estrogen receptor-positive human breast cancer cells (MCF-7). Comparative studies in the normal human skin FC and liver HC cell lines showed that the beetroot extract had significantly lower cytotoxic effect than doxorubicin. The results suggest that betanin, the major betacyanin constituent, may play an important role in the cytotoxicity exhibited by the red beetroot extract.

The in vitro inhibitory effect of Beta vulgaris (beet) root extract on Epstein-Barr virus early antigen (EBV-EA) induction using Raji cells revealed a high order of activity compared to capsanthin, cranberry, red onion skin and short and long red bell peppers. An in-vivo anti-tumor promoting activity evaluation against the mice skin and lung bioassays also revealed a significant tumor inhibitory effect. The combined findings suggest that beetroot ingestion can be one of the useful means to prevent cancer. (Kapadia et al., 1996).

References

Georgiev VG, Weber J, Kneschke EM, Denev PN, Bley T, Pavlov AI. Antioxidant activity and phenolic content of betalain extracts from intact plants and hairy root cultures of the red beetroot Beta vulgaris cv. Detroit dark red. Plant Foods Hum Nutr. 2010 Jun;65(2):105-11.

Kapadia GJ, Azuine MA, Rao GS, Arai T, Iida A, Tokuda H. Cytotoxic effect of the red beetroot (Beta vulgaris L.) extract compared to doxorubicin (Adriamycin) in the human prostate (PC-3) and breast (MCF-7) cancer cell lines. Anticancer Agents Med Chem. 2011 Mar;11(3):280-4.

Kapadia GJ, Tokuda H, Konoshima T, Nishino H. Chemoprevention of lung and skin cancer by Beta vulgaris (beet) root extract. Cancer Lett. 1996 Feb 27;100(1-2):211-4.

Kapil V, Milsom AB, Okorie M, Maleki-Toyserkani S, Akram F, Rehman F, Arghandawi S, Pearl V, Benjamin N, Loukogeorgakis S, Macallister R, Hobbs AJ, Webb AJ, Ahluwalia A. Inorganic nitrate supplementation lowers blood pressure in humans: role for nitrite-derived NO. Hypertension. 2010 Aug;56(2):274-81. Epub 2010 Jun 28.

Kenjale AA, Ham KL, Stabler T, Robbins JL, Johnson JL, Vanbruggen M, Privette G, Yim E, Kraus WE, Allen JD. Dietary nitrate supplementation enhances exercise performance in peripheral arterial disease. J Appl Physiol. 2011 Jun;110(6):1582-91. Epub 2011 Mar 31.

Rokkedal-Lausch, T, Franch, J., Poulsen, M.K., Thomsen, L.P., Weitzberg, E., Kamavuako, E.N., Karbing, D.S., and Larsen, R.G. 2019. Chronic high-dose beetroot juice supplementation improves time trial performance of well-trained cyclists in normoxia and hypoxia. Nitric Oxide, (2019), 85:44-52. Doi: 10.1016/j.niox.2019.01.011.

5.3. Radish (*Raphanus stivus* L.)

The radish (*Raphanus sativus* L.) is a common edible root vegetable belonging to the Brassicaceae cultivated and consumed world wide. Radishes have numerous varieties, varying in size, color and duration required for cultivation. There are some radishes that are grown for their seeds; oilseed radishes are grown, as the name implies, for oil production. In India radish is mostly used as salad and vegetable for cooking. It is also used for making Indian bread or parathas. Radishes are rich in ascorbic acid, folic acid, and potassium. They are a good source of vitamin B6, riboflavin, magnesium, copper, and calcium.

Cultivars/Varieties: The popular asiatic varieties under cultivation are Pusa Deshi, Pusa Chetki, Pusa Reshmi, Japanese White, Punjab Safed, Punjab Pasand, Punjab Agethi, Kalyanpur No-1, Arka Nishant and Co-1. The European varieties are Pusa Himani, White Icicle, Scarlet Globe, Scarlet Long and Kashi Sweta.

Nutritional Components of Radish

Constituents	Quantity (per 100 g)	Constituents	Quantity (per 100 g)
Carbohydrates	3.40 g	Iron	0.34 mg
Proteins	0.68 g	Magnesium	10 mg
Fats	0.10 g	Thiamine	0.012 mg
Dietary Fibre	1.6 g	Riboflavin	0.039 mg
Energy	16 Kcal	Niacin	0.254 mg
Phosphorus	20 mg	Pantothenic acid	0.165 mg
Potassium	233 mg	Folic Acid	25 ug
Zinc	0.28 mg	Vitamin C	14.8 mg
Calcium	25 mg	Vitamin B6	0.071

Source: (USDA Nutrient Database)

Phytochemicals

Throughout history radishes have been effective when used as a medicinal food for liver disorders. They contain a variety of sulfur-based chemicals that increase the flow of bile. Therefore, they help to maintain a healthy gallbladder and liver, and improve digestion. Fresh radish roots contain a larger amount of vitamin C than cooked radish roots. Radish greens contain far more vitamin C, calcium, and protein than the roots. The 2-primary radish glucosinolates i.e., glucoraphasatin, and glucoraphanin, were isolated using solid phase extraction followed by preparative HPLC purification (Scholl *et al.*, 2011). Raphasatin, the isothiocyanate metabolite of glucoraphasatin and the oxidized counterpart of sulphoraphene, is a highly unstable compound. Despite the instability of raphasatin, dietary exposure to radishes produced significant induction of detoxification enzymes.

Alkaloid and nitrogen compounds present in the roots are pyrrolidine, phenethylamine, N-methylphenethylamine, 1,2′-pyrrolidin-tion-3-il-3-acid-carboxilic-1,2,3,4-tetrahydro-β-carboline, and sinapine (Marquardt, 1976; Wan,

1984; Weilan *et al.*, 1987). Aesculetin and scopoletin were the two hydroxycoumarins identified from Radish (Stoehr and Hermann,1975). B-fructosidase, cysteine synthase, β-galactosidase, hydroxycinnamoyl transferase, catalase, glutathione reductase and γ-glutamyl transpeptidase are some of the very important enzymes found in Radish roots. All the members of the family Braissicaceae are rich in glucosinolates. Glucosinolates are very stable water-soluble precursors of isothiocyanates. Four major organic acids are present in the roots of the radish: oxalic, malic, malonic, and erythorbic acid. Radishes and horseradish showed caffeic, *p*-coumaric, ferulic, hydroxycinnamic, *p*-hydroxybenzoic, vanillic, salicylic, and gentisic acid (Stoehr and Hermann, 1975). Among the anthocyanins, pelargonidine and cyanidine were responsible for red and violet color in corollas and roots of coloured radishes. The absence of pelargonidine and cyanidine resulted in a white color. The major anthocyanins of radishes are pelargonidin-3-sophoroside-5-glucoside acetylated with malonic acid and either ferulic or p-coumaric acid. The proteoglycan consiste of 86% of a polysaccharide component-contained L-arabinose and D-galactose as major sugar constituents, together with small proportions of D-xylose, D-glucose, and uronic acids, and 9% of a hydroxyproline-contained protein. Radish leaves contain only one of the sulfonium diatero isomers of S-adenosyl methionine (AdoMet), which has a remarkable variety of biochemical functions. It is an allosteric enzyme effector and a precursor of spermine biosynthesis, spermidine, and ethylene. It is also the methyl group donor for most biological transmethylation reactions,

Health Benefits of Radish based on Traditional/Folklore Practices

Radish is good as appetizer, mouth and breathe freshener, laxative, regulates metabolism, improves blood circulation, is a good treatment for headache, acidity, constipation, nausea, obesity, sore throat, whooping cough, gastric problems, gallbladder stones, dyspepsia *etc.* It has got several uses in traditional home remedies. The benefits of radish against certain ailments and on certain body parts are listed below:

Jaundice: Radish is very good for the liver and the stomach and it is a very good detoxifier too. It is a miracle food useful in jaundice as it helps removing bilirubin and checks its production. It also checks destruction of red blood cells during jaundice by increasing supply of fresh oxygen in the blood. The black radish is more preferred in jaundice. The leaves of radish are also very useful in treatment of jaundice.

Piles: Radish is very rich in roughage, i.e. indigestible carbohydrates. This facilitates digestion, retains water, cures constipation (one of the main causes for piles) and thus gives relief in piles. Being a very good detoxifier, it helps heal up piles fast. Its juice also soothes the digestive and excretory system.

Urinary Disorders: Radishes are diurectic in nature, i.e. increase production of urine. Juice of radish also cures inflammation and burning feeling during urinating. It also cleans the kidneys and inhibits infections in kidneys and urinary system. Thus, it helps a great deal in curing urinary disorders.

Weight Loss: Radishes are very filling, i.e. fills your stomach and satisfies the hunger easily without giving many calories, as they are low in digestible carbohydrates, high in roughage and contain a lot of water. It is a very good dietary option for those determined to lose weight.

Cancer: Being a very good detoxifier and rich in vitamin-C, folic avid and anthocyanins, radish helps cure many types of cancer, particularly those of colon, kidney, intestines, stomach and oral cancer.

Leucoderma: The detoxifying and anti carcinogenic properties of radish make it useful in treatment of aphanina. The radish seeds are traditionally used in treating leucoderema. The radish seeds should be powdered and soaked in vinegar or ginger juice or cow's urine and then applied on the white patches. Eating radish also aids cure of aphanina.

Skin Disorders: Vitamin-C, phosphorus, zinc and some members of vitamin-B complex, which are present in radish, are good for skin. The water in it helps maintaining moisture of the skin. Smashed raw radish is a very good cleanser and serves as a very efficient face pack. Due to its disinfectant properties, radish also helps cure skin disorders, such as drying up, rashes, cracks *etc.*

Insect Bites: It has anti pruritic properties and can be used as an effective treatment for insect bites, stings of bees, hornets, wasps *etc.* Its juice also reduces pain and swelling and soothes the affected area.

Fever: It brings down the body temperature and relieves inflammation due to fever. Drinking radish juice mixed with black salt is said to bring down the temperature. Being a good disinfectant, it also fights infections which cause fever.

Respiratory Disorders, Bronchitis and Asthma: Radish is an anti-congestive, i.e. it relieves congestion of respiratory system including nose, throat, windpipe and lungs, due to cold, infection, allergies and other causes. It is a good disinfectant and rich in vitamins, which protect respiratory system from infections.

Liver & Gallbladder: Radish is especially beneficial for liver and gallbladder functions. It regulates production and flow of bile and bilirubin, acids, enzymes and removes excess bilirubin from the blood, being a good detoxifier. It also contains enzymes like myrosinase, diastase, amylase and esterase. It protects liver and gallbladder from infections and ulcers and soothes them.

Therapeutic Properties of Radish

Antimicrobial Activity: Radish has antibacterial and antifungical activities. Studies showed that crude juice of the radish inhibited the growth of *Escherichia coli, Pseudomonas pyocyaneus, Salmonella typhi,* and *Bacillus subtilis in vitro.* Caffeic acid showed antifungal properties *in vitro* against *Helminthosporium maydis.* The highest fungicidal activity depended on concentration of isothiocyanates (Smolinoka and Horbowicz, 1999).

Antioxidant Activity: The red radish pigment (pelargodinin-3-sophoroside-5-glucoside) had almost the same antioxidative activity as BHT at the same concentration.

Antitumor Activity: A neutral fraction of kaiware radish aqueous extract, *in vitro* showed inhibition of proliferation of mouse embryo fribroblast 3T3 cells and papovavirus SV40 transformed 3T3 cells (Akihiro *et al.*, 1999), which confirm its antitumor activity.

Antiviral Activity: Caffeic acid and pelargonidin present in radishes were found to be virucidal for several enveloped viruses (Strack *et al.*, 1985). The lipopolysaccharides of radish showed activity against Herpes virus (antiherpes).

Platelet Aggregation Inhibitor

The 6-methyl-sulfinylhexyl-isothiocyanate (MS-ITC) was isolated from wasabi horseradish (Japanese domestic) as a potential inhibitor of human platelet aggregation *in vitro.*

Cardiovascular Disease Prevention

In trials conducted on animal models, radish powder decreased the lipid levels by increasing the fecal excretion of total lipids, triglycerides, and total cholesterol. Catalase and glutathione peroxidase (GSH-Px) activities in red blood cell (RBC) were most remarkably increased by radish. Superoxide dismutase (SOD), catalase, and GSH-Px activities in the liver were increased by radish powder. Xanthine oxidase (XOD) activities in the liver were decreased by radish. Flavonoids and vitamin C in radish were found to inhibit lipid peroxidation, promote liver and RBC catalase, and inhibit XOD activities in animals' tissues (Jin and Kyung, 2001).

Cancer Prevention

Epidemiological studies give evidence that cruciferous vegetables (CF) confer protection against cancers. Results from animal experiments show that they reduce chemically induced tumor formation. These properties have been attributed to alterations in the metabolism of carcinogens by breakdown products of glucosinolates, which are constituents of CF (Steinkellner *et al.*, 2001). The scientists carried out crossover intervention studies and found pronounced GST-induction upon consumption of Brussels sprouts and red cabbage. Furthermore, they found that the isoenzyme induced was GST-pi which plays an important role in protection against breast, bladder, colon and testicular cancer. Kim *et al.*, (2011) investigated the effects of the ethanol extract of aerial parts of *Raphanus sativus* L. (ERL) on breast cancer cell proliferation and gene expression associated with cell proliferation and apoptosis in MDA-MB-231 human breast cancer cells and suggested that *Raphanus sativus*, L. inhibits cell proliferation via the ErbB-Akt pathway in MDA-MB-231 cells.

Detoxification of Liver

Hanlon *et al*., (2007) showed that crude aqueous extract from 0.3 to 3 mg of dry SBR (Spanish Black Radishes) increased the activity of the phase II detoxification enzyme quinone reductase in the human hepatoma HepG2 cell line with a maximal effect at a concentration of 1 mg/mL. They also showed that the isothiocyanate metabolite of glucoraphasatin, 4-methylthio-3-butenyl isothiocyanate (MIBITC), significantly induced phase II detoxification enzymes at a concentration of 10 microM. These data demonstrate that the crude aqueous extract of Spanish Black Radish (SBR) and the isothiocyanate metabolite of glucoraphasatin, *i.e.*, MIBITC, are potent inducers of detoxification enzymes in the HepG2 cell line.

References

Akihiro, M., Koji, K., Hiroyoshi, O., Kazuaki, K., and Yoshiko, A. 1999. Antitumor substances from vegetables, their manufacture, and pharmaceutical compositions. Jpn. Kokai Tokkyo Koho JP 11 49,793[99 49,793] (Cl. C07G17/00), 23 Feb 1999, Appl. 97/215224, 8 Aug 1997.

Hanlon, P.R., Webber, D.M., and Barnes, D.M. 2007. Aqueous extract from Spanish black radish (*Raphanus sativus* L. var. *niger*) induces detoxification enzymes in the HepG2 human hepatoma cell line. J Agric Food Chem., (2007), 55(16):6439-46.

http://www.everynutrient.com/healthbenefitsofradishes.html

http://www.greenlifepages.com/body-mind/item/439-the-surprising-benefits-of-radishes.

http://www.nutrition-and-you.com/radish.html

Jin, A.S. and Kyung, K.M. 2001. Effect of dry powders, ethanol extracts and juices of radish and onion on lipid metabolism and antioxidative capacity in rats. *Han'guk Yongyanghak Hoeji* 34: 513–524.

Kim, W.K., Kim, J.H., Jeong, da H., Chun, Y.H., Kim S.H., Cho, K.J., Chang, M.J. 2011. Radish (*Raphanus sativus* L. leaf) ethanol extract inhibits protein and mRNA expression of ErbB(2) and ErbB(3) in MDA-MB-231 human breast cancer cells. Nutr Res Pract., (2011), 5(4):288-93.

Marquardt, P.1976. N-methylphenethylamine in vegetables. *Arzneimittel forschung* 26: 201-203.

Scholl, C., Eshelman, B.D., Barnes, D.M., Hanlon, P.R. 2011. Raphasatin is a more potent inducer of the detoxification enzymes than its degradation products. J Food Sci., (2011), 76(3):C504-11. Doi: 10.1111/j.1750-3841.2011.02078.x.

Smolinska, U., and Horbowicz, M. 1999. Fungicidal activity of volatiles from selected cruciferous plants against resting propagules of soil-borne fungal pathogens. *J. Phytopathol.* 147: 119–124.

Steinkellner, H., Rabot, S., Freywald, C., Nobis, E., Scharf, G., Chabicovsky, M., Knasmüller, S., and Kassie, F. 2001. Effects of cruciferous vegetables and their constituents on drug metabolizing enzymes involved in the bioactivation of DNA-reactive dietary carcinogens. Mutat Res., (2001), 1:480-481:285-97.

Stoehr, H. and Herrmann, K. 1975. Phenolic acids of vegetables. III. Hydroxycinnamic and hydroxybenzoic acids of root vegetables. *Z.* Lebensm. Unters. Forsch., 159: 219–224.

Strack, D., Pieroth, M., Scharf, H., and Sharma, V. 1985. Tissue distribution of phenylpropanoid metabolism in cotyledons of *Raphanus sativus* L. *Planta* 164: 507–511.

Wan, C. 1984. Studies on chemical constituents in radish (*Raphanus sativus* L.) seeds. II. *Shaanxi Xinyiyao*, 13: 54–55.

Weilan, W., Jin, Z., Zhongda, L., and Meng, L. 1987. Hypotensive constituents of Laifuzi (*Semen aphanin*). *Zhongcaoyao* 18: 101–103.

5.4. Turnip (*Brassica compestris var. rapa* L.)

Turnip (Brasicca *rapa* L.Syn. *Brassica campestris* var. *rapa*) belongs to the family Brassicaceae. It is a thick underground portion of hypocotyl which may vary in shape from flat to globular to top-shaped and long. The colour may vary from while, or yellow to red, purple, and green. It is popular winter vegetable successfully grown at an elevation of 1500 m above MSL. Turnip is a popular root vegetable in North India and is used as salad, cooked or pickled. Turnip tops are extensively used as cattle feed.

Like radish, in turnips also the two types like Tropical or Asiatic and Temperate or European. Asiatic types are pungent and are generally used for making pickles while European types are relatively sweeter and palatable as salads.

Cultivars/Varieties: Based on the morphology of root and top, there are two varieties. In white fleshed varieties, top may be purple, green or white and in yellow fleshed varieties the top is purple or yellow. The European varieties grown in India are Pusa Chandrima, Pusa Swarnima, Golden Ball, Purple Top White Globe and Snowball. The popular Asiatic varieties are Pusa Sweti, Pusa Kanchan and Turnip L-1.

Uses

Most baby turnips can be eaten whole, including their leaves. Their flavor is mild, so they can be eaten raw in salads like radish and other vegetables. In India, particularly in northern parts, it is widely used in pickling. Turnip is rich is dietary fibre and consumed for its mild pungent odour. Turnips are very low-calorie root vegetables; contains only 22 Kcalories per 100 g. However, they are very good source of antioxidants, minerals, vitamins and dietary fiber.

Nutritional Components of Turnip Roots

Constituent	Quantity (per 100g)	Constituent	Quantity (per 100g)
Moisture		Sodium	16 mg
Carbohydrates	6.43 g	Zinc	0.12 mg
Proteins	0.90 g	Thiamine	0.027 mg
Fats	0.10 g	Riboflavin	0.023 mg
Dietary Fibre	3.50 g	Niacin	0.299 mg
Calcium	33 mg	Pantothenic acid	0.142mg
Iron	0.18 mg	Vitamin B6	0.067mg
Magnesium	9 mg	Folate	9 µg
Phosphorus	26 mg	Vitamin C	11.6 mg
Potassium	177 mg	Energy	22 Kcal

Gopalan et al., (1985)

Phytochemicals

Hydroxycinnamic acid, quercetin, myricetin, isorhamnetin, and kaempferol are among the key antioxidant phytonutrients provided by turnip greens. This broad-spectrum antioxidant support helps lower the risk of oxidative stress in our cells. Two key glucosinolates that have been clearly identified in turnip greens in significant amounts are gluconasturtian and glucotropaeolin.

Health Benefits of Turnip roots and greens

Unlike other cruciferous vegetables, turnip greens have not been the direct focus of most health-oriented research studies. However, turnip greens have sometimes been included in a longer list of cruciferous vegetables that have been clubbed together and studied to determine potential types of health benefits. Based upon several dozen studies involving cruciferous vegetables as a group (and including turnip greens on the list of vegetables studied), cancer prevention appears to be a standout area for turnip greens when summarizing health benefits.

- Because it has a high content of vitamin C, and other antioxidants, turnips are a powerful anti-inflammatory. It also prevents the development of asthma symptoms.
- Eating turnips will aid the body in fighting against the development or progression of cholesterol buildup.
- It is believed that turnip consumption also inhibits joint damage, risk of osteoporosis and the incidence of Rheumatoid Arthritis.The turnip, besides having calcium, is also an excellent source of copper, a mineral necessary in the body's production of connective tissues.
- They are a good source of folic acid, manganese, thiamine, magnesium, riboflavin among others and is said to help in proper functioning of the body's immune system.
- **Detoxing**: Turnip greens contain phytonutrients called glucosinolates that can help activate detoxification enzymes and regulate their activity. Two key glucosinolates that have been clearly identified in turnip greens in significant amounts are gluconasturtian and glucotropaeolin.
- **Antioxidant**: As an excellent source of vitamin C, vitamin E, beta-carotene, and manganese, turnip greens provide highest level support for four conventional antioxidant nutrients. Hydroxycinnamic acid, quercetin, myricetin, isorhamnetin, and kaempferol are among the key antioxidant phytonutrients provided by turnip greens. By providing a diverse array of antioxidant nutrients, turnip greens help to lower the cancer risk.
- **Anti-inflammatory Benefits**: As an excellent source of vitamin K and a good source of omega-3 fatty acids (in the form of alpha-linolenic acid, or ALA), turnip provide two hallmark anti-inflammatory nutrients. Vitamin K acts as a direct regulator of the inflammatory response, and ALA is the building block for several of the body's most widely used families of anti-inflammatory messaging molecules. Like chronic oxidative stress and chronic weakened detox ability, chronic unwanted inflammation can significantly increase our risk of cancers and other chronic diseases (especially cardiovascular diseases).
- **Cardiovascular Benefits**: Researchers have looked at a variety of cardiovascular problems—including heart attack, ischemic heart disease, and atherosclerosis—and found preliminary evidence of an ability on the part of cruciferous vegetables to lower the risk of these health problems.

6. Tuber Crops

6.1. Potato (*Solanum tuberosum* L.)

Potato, *Solanum tuberosum,* is a starchy, tuberous crop belonging to the Solanaceae family. The name potato originally referred to a type of sweet potato, *batata,* in Spanish. Potatoes are occasionally referred to as "Irish potatoes" or "white potatoes" in the United States, to distinguish them from sweet potatoes. The potato was believed to be first domesticated in the region of southern Peru and extreme northwestern Bolivia between 8000 to 5000 BC. The potato diffused widely after 1600, becoming a major food resource in Europe and East Asia. Greater geographic mobility of farmers led to the rapid spread of potato cultivation throughout China, and India. China is now the world's largest potato producer and nearly one third of the world's potatoes are produced in China and India.

Today potato is the fourth major food crop after rice, wheat and maize in the world. In India potato is a very important crop and are mainly produced in the states of Uttar

Cultivars/Varieties: Most of the improved potato varieties were developed by CPRI, Kufri, Shimla. Most of the earlier varieties released as clonal selections derived from foreign varieties have been replaced with high yielding new varieties from the same institute. The popular varieties under cultivation now are Kufri Sindhuri, Kufri Chandramukhi, Kufri Jyothi, Kufri Lauvkar, Kufri Muthu, Kufri Dewa, Kufri Badshah, Kufri Bahar, Kufri Lalima, Kufri Swarna, Kufri Jawahar, Kufri Ashoka, Kufri Megha, Kufri Sutlej, Kufri Pukhraj, Kufri Chipsona-1, Kufri Chipsona-2 and Kufri Giriraj.

Uses

Potatoes are used in different ways like boiled, fried, baked and roasted. Potatoes are used to brew alcoholic beverages such as vodka, potcheen, or akvavit.they are also used as food for domestic animals. Potato starch is used in the food industry as, for example, thickeners and binders of soups and sauces, in the textile industry, as adhesives, and for the manufacturing of papers and boards. Many companies are exploring the possibilities of using waste potatoes to obtain polylactic acid for use in bioplastic products; other research projects seek ways to use the starch as a base for biodegradable packaging. Potato skins, along with honey, are used in folk medicine for burns in India. Burn centers in India have experimented with the use of the thin outer skin layer to protect burns while healing. The potato is best known for its carbohydrate content, which is in the form of starch. Because of its high starch content, it is a staple food in several parts of world including in some parts of India. A small but significant portion of this starch is resistant to digestion by enzymes in the stomach and small intestine, and so reaches the large intestine essentially intact, thereby, increasing its physiological significance. Besides these the potato also provides various other nutritional constituents to the body. Potato is also processed into various products like chips, powder, pappad, savory items.

Nutirtional Components of Potato

Constituents	Quantity (per 100 g)	Constituents	Quantity (per 100 g)
Moisture	79 g	Potassium	421 mg
Carbohydrates	17 g	Sodium	6 mg
Proteins	2.0 g	Zinc	0.29 mg
Fats	0.09 g	Copper	0.11 mg
Fibers	2.20 g	Manganese	0.15 mg
Energy		Vitamin C	19.7 mg
Calcium	12 mg	Thiamine	0.08 mg
Iron	0.78 mg	Niacin	0.05 mg
Magnesium	23 mg	Folate	16 mcg
Phosphorus	57 mg	Pantothenic acid	0.3 mg
Selenium	0.3 mg	Vitamin E	0.01 mg

Gopalan *et al.*, (1985)

Phytochemicals

Potato contains several phytochemicals such as phenolics, flavonoids, polyamines, and carotenoids, which are highly desirable in diet because of their beneficial effects on human health. Phenolic compounds represent a large group of minor chemical constituents in potatoes, which play an important role in determining their organoleptic properties. Chlorogenic acid is one of the phenols which constitutes up to 90% of the phenols in potato. Others found in potatoes are 4-O-caffeoylquinic acid (crypto-chlorogenic acid), 5-O-caffeoylquinic (neo-chlorogenic acid), 3,4-dicaffeoylquinic and 3,5-dicaffeoylquinic acids. The potato contains vitamins and minerals like thiamin, riboflavin, folate, niacin, magnesium, phosphorus, iron, and zinc.

A study by the USDA has indicated that there is considerable variation in phytochemical content between potato varieties, and that some varieties accumulate significant levels of bioactive compounds such as phenolic compounds including flavonoids, quercetin and kukoamines. Such compounds have been implicated in reduction of risk for cardiovascular diseases, high blood pressure and certain cancers.

Health Benefits of Potato

Weight Gain: Potatoes mainly constitutes carbohydrates, hence are ideal diet for putting on weight. The vitamins like vitamin-C and B-complex also help in proper absorption of this carbohydrate by the body. That is why they make an inevitable part of the diet of Sumo Wrestlers.

Digestion: Since potatoes predominantly contain carbohydrates, they are easy to digest and ready source of energy. This property makes them a good diet for patients, babies and those who cannot digest hard food but need energy. But eating too much of potatoes regularly may cause acidity and flatulence. Potatoes also contain considerable amount of fiber or roughage too, particularly in raw potatoes.

Skin Care: Vitamin-C and B-complex and minerals like potassium, magnesium, phosphorus and zinc are good for the skin. Apart from that, pulp obtained from crushed raw potatoes, mixed with honey, can serve as excellent skin and face packs. This helps even curing pimples and spots on the skin. The pulp, if applied externally on burns, gives a quick relief and heals fast. Smashed potatoes, even water in which potatoes are washed, are very good for softening and cleaning skin, especially around elbows, back of the palms *etc.*

Rheumatism: Vitamins, calcium and magnesium in potatoes is believed to give relief from rheumatism. Water obtained from boiling potato is more useful for this purpose, however, it tends to increase body weight which may have adverse effects on rheumatic people.

Inflammation: Potato is very effective in inflammation, internal or external. Since it is soft, easily digestible and has a lot of vitamin-C (very good anti-oxidant and repairs wears and tears), potassium and vitamin-B6, among others, it relieves inflammation of intestines and the digestive system. It is very good diet for those who have mouth ulcers. Raw smashed potato can be applied to external inflammation, burns *etc.* to get relief.

High Blood Pressure: Potato can help in managing high blood pressure due to tension, indigestion *etc.* due to abundance of vitamin-C and B in it. But it should be avoided if the person is diabetic.The fiber present in potato is helpful in lowering cholesterol and improves functioning of insulin in the body.

Brain Function: Proper functioning of the brain depends largely on the glucose level, oxygen supply, some members of the vitamin-B complex and some hormones, amino acids and fatty acids like omega-3 fatty acids. Potatoes cater to almost all the needs mentioned above. They are high in carbohydrates and thus maintain good level of glucose in the blood which does not let brain fatigue creep in and keeps the brain active and alert. Oxygen is carried to the brain by the haemoglobin in the blood and whose main constituent is iron. Potato contains iron too and thus aids in this function. Potato is rich in vitamin-B6 and contains traces of other members of vitamin B-complex. In addition, it contains certain other elements like phosphorus and zinc which are good for brain too.

Heart Diseases: Apart from the vitamins (B-complex, C), minerals and roughage, potatoes also contain certain substances called carotenoids (lutein, zeaxanthin *etc.*) which are beneficial for heart .

Diarrhea: It is an excellent energy-rich diet for those suffering from diarrhea, since it is very easy to digest.

Other Benefits and Cautions: Juice of potato is a good treatment for burns, bruises, sprains, skin problems, ulcers, effect of narcotics, cancer of prostrate and uterus and formation of cysts or tumors. On the other hand, some care also needs to be taken while eating potatoes. Green potatoes are poisonous, and so are potato leaves and fruits, as they contain alkaloids like solanine, chaconine and arsenic whose overdose may prove fatal. Moreover, the glycemic index (in simple words, the energy or sugar content) of potatoes is very high (above 80) and so obese, diabetic and those who are planning to reduce the weight should avoid eating potatoes.

Therapeutic Properties of Potato

Incomplete digestion of dietary starch in small intestinal has received increased attention in recent years (Calvert et al., 1989). Several reports have demonstrated that incomplete small intestinal hydrolysis of starch may be a normal phenomenon in humans (Levitt, 1983), allowing a significant quantity of starch to enter the colon. It has been recently suggested that this undigested starch may have some role in providing effects previously ascribed to dietary fiber, such as enhancement of fecal weight (Shetty and Kurpada, 1986) or even a role in colon cancer prevention (Thornton et al., 1987).

Studies conducted by Calvert et al., (1989) showed that resistant potato starch (RPS), in contrast to cooked potato starch (CPS), increases fecal weight, slows gastrointestinal transit time, and increases colon mucosal cell proliferation. They attributed these effects to the entry of significant quantities of undigested starch (resistant starch) into the large intestine. Numerous previous investigators (Fleming and Vose, 1979; Booher et al., 1951; Baker et al., 1950; Reussner et al., 1963) have documented the marked resistance of RPS in small intestinal digestion, in contrast to the much more digestible CPS.

The effect of resistant starch (RS) on postprandial plasma concentrations of glucose, lipids, and hormones, and on subjective satiety and palatability ratings was investigated in 10 healthy, normal-weight, young males by Raben et al., (1994). They found that, the replacement of digestible starch with RS resulted in significant reductions in postprandial glycemia and insulinemia, and in the subjective sensations of satiety.

Starches of different digestibilities may enter the colon to different extents and alter colonic function. Lajvardi et al., (1993) fed male Fischer 344 rats with diets containing 25% cooked potato starch, arrowroot starch, high amylose cornstarch or raw potato starch for 6 wk. They found that raw potato starch was much less digested than high amylose cornstarch, resulting in a 32-fold greater amount of undigested starch entering the cecum. Entry of a large amount of raw potato starch into the colon resulted in greater luminal acidity, greater luminal bulk and slower transit. A much smaller amount of starch entered the colon in the high amylose cornstarch group and resulted in fecal bulking but no alteration in transit.

The potential effect of a long-term intake of resistant starch on colonic fermentation and on gut morphologic and immunologic indices of interest in bowel conditions was studied in pigs by Nofrarias et al., (2007). They suggested that long-term intake of RPS induces pronounced changes in the colonic environment, reduces damage to colonocytes, and improves mucosal integrity, reducing colonic and systemic immune reactivity, to which health benefits in inflammatory conditions are associated.

Some alkaloids and antioxidants in fruits and vegetable were also found to reduce several chronic diseases. α-Solanine, a naturally occurring steroidal glycoalkaloid in potato sprouts, was found to possess anti-carcinogenic properties, such as inhibiting proliferation and inducing apoptosis of tumor cells. The studies of Lu *et al.*, (2010) suggested that α-solanine inhibited migration and invasion of A2058 cells by reducing MMP-2/9 activities. It also inhibited JNK and PI3K/Akt signaling pathways as well as NF-κB activity. This study reveals a new therapeutic potential for α-solanine in anti-metastatic therapy.

Polyphenols from fruits and vegetables exhibit anticancer properties both *in vitro* and *in vivo* and specialty potatoes are an excellent source of dietary polyphenols, including phenolic acids and anthocyanins. Reddivari *et al.*, (2007) investigated the effects of specialty potato phenolics and their fractions on LNCaP (androgen dependent) and PC-3 (androgen independent) prostate cancer cells. Their studies showed potato extract and AF induced apoptosis in both the cells. It was reported for the first time that cytotoxic activities of potato extract/AF in cancer cells were due to activation of caspase-independent apoptosis. Protein as well as starch is fermented in the colon, but the interaction between protein and starch fermentation and the impact on colonic oncogenesis was unknown till recently. Le Leu *et al.*, (2006) investigated the interaction of resistant starch (RS) with digestion of resistant potato protein (PP) on colonic fermentation events and their relationship to intestinal tumourigenesis. They were of the opinion that RS altered the colonic luminal environment by increasing the concentration of short-chain fatty acid including butyrate and lowering production of potentially toxic protein fermentation products. These effects of RS not only protected against intestinal tumourigenesis but also ameliorated the tumour-enhancing effects of feeding indigestible protein.

Epidermal growth factor (EGF) and its receptor (EGFR) are involved in many aspects of the development of carcinomas, including tumor cell growth, vascularization, invasiveness, and metastasis. Because EGFR has been found to be overexpressed in many tumors of epithelial origin, it is a potential target for antitumor therapy. Blanco *et al.*, (1998) reported that potato carboxypeptidase inhibitor (PCI), a 39-amino acid protease inhibitor with three disulfide bridges, is an antagonist of human EGF. It competed with EGF for binding to EGFR and inhibited EGFR activation and cell proliferation induced by this growth factor. PCI suppressed the growth of several human pancreatic adenocarcinoma cell lines, both *in vitro* and *in vivo* (mice). PCI has a special disulfide scaffold called a 'T-knot' that is also present in several growth factors including EGF and transforming growth factor alpha.

References

Baker, F., Nasr, H., Morrice, F. and Iabez, B. 1950. Bacterial breakdown of structural starches and starch products in the digestive tract of ruminant and non-ruminant mammals. Pathol Bacterial., 62: 617-638.

Blanco-Aparicio, C., Molina, M.A., Fernández-Salas, E., Frazier, M.L., Mas, J.M., Querol, E., Avilés, F.X., and de Llorens, R. 1998. Potato carboxypeptidase inhibitor, a T-knot protein, is an epidermal growth factor antagonist that inhibits tumor cell growth. J Biol Chem., (1998), 273(20):12370-7.

Booher,L. E., Behan,I., and McMeans, E. 1951. Biologic utilizations of unmodified and modified food starches. Nutr. 45:75-95.

Calvert Richard, J., Otsuka, M., and Satchithanandam, S.1989. Consumption of raw potato starch alters intestinal function and colonie cell proliferation in the rat. J. Nutr., 119 (11): 1610-1616, 1989.

Fleming,S. E., and Vose,J. R. 1979. Digestibility of raw and cooked starches from legume seeds using the laboratory rat. Nutr., 109: 2067-2075.

http://en.wikipedia.org/wiki/Potato

http://www.ipfn.ie/research/associatedwalsh/profilingphytoc.html

http://www.organicfacts.net/health-benefits/vegetable/health-benefits-of-potato.html

http://www.potatogoodness.com/

http://www.thehotpotato.com/english/potato_facts.htm

Lajvardi, A., Mazarin, G.I., Gillespie, M.B., Satchithanandam, S., and Calvert, R.J. 1993. Starches of varied digestibilities differentially modify intestinal function in rats. J Nutr., (1993), 123(12):2059-66.

Le Leu, R.K., Brown, I.L., Hu, Y., Morita, T., Esterman, A., Young, G.P. 2006. Effect of dietary resistant starch and protein on colonic fermentation and intestinal tumourigenesis in rats. Carcinogenesis, (2007), 28(2):240-5.

Levitt, M. D. 1983. Malabsorption of starch: a normal phenomenon. Gastroenterology, (1983), 85: 769-770.

Lu, M.K., Shih, Y.W., Chang, C.T.T., Fang, L.H., Huang, HC., and Chen, P.S. 2010.α-Solanine inhibits human melanoma cell migration and invasion by reducing matrix metalloproteinase-2/9 activities. Biol Pharm Bull., (2010), 33(10):1685-91.

Nofrarías, M., Martínez-Puig, D., Pujols, J., Majó, N., and Pérez, J.F. 2007. Long-term intake of resistant starch improves colonic mucosal integrity and reduces gut apoptosis and blood immune cells. Nutrition, (2007), 23(11-12):861-70.

Raben, A., Tagliabue, A., Christensen, N.J., Madsen, J., Holst, J.J., and Astrup, A. 1994. Resistant starch: the effect on postprandial glycemia, hormonal response, and satiety. Am J Clin Nutr., (1994), 60(4):544-51.

Reddivari, L., Vanamala, J., Chintharlapalli, S., Safe, S.H., and Miller, J.C, Jr. 2007. Anthocyanin fraction from potato extracts is cytotoxic to prostate cancer cells through activation of caspase-dependent and caspase-independent pathways. Carcinogenesis, (2007), ;28(10):2227-35.

Reussner, G., Andros, J. and Theissen, J. R. R. 1963. Studies on the utilization of various starches and sugars in the rat. Nutr. 80: 291-298.

Shetty,P. S., and Kurpad,A. V. 1986. Increasing starch intake in the human diet increases fecal bulking. Am. Clin. Nutr. 43: 210-212.

Stephena, M., Haddada. C., and Phillips, S, F. 1983. Passage of carbohydrate into the colon. Direct measurements in humans. Gastroenterology, 85: 589-595.

Thornton,I. R., Dryden,A., Kelleher, J,. and Losowsky, M, . S. 1987. Super-efficient starch absorptionâ€" a risk factor for coIonic neoplasia? Dig. Dis. Sci., 32: 1088-1091.

6.2. Sweet Potato (*Ipomoea batatas* L.)

The sweet potato belongs to the family Convolvulaceae and is herbaceous perennial but cultivated as annual. In India it is popularly known as 'Sakarkand'. Sweet potato is native to tropical America. It is an important tuber crop in tropical and subtropical countries like Africa, China and India. In India, grown in an area of 1.13 lakh hectares with a total production of 1.05 million tonnes. It is mainly produced in Orissa, West Bengal, Uttar Pradesh, Assam, Chattisgarh and Karnataka. The average productivity is 9.3 tonnes/ha.

Cultivars/Varieties: Sweet potato varieties differ in shape, size and colour of leaves, tubers and colour of flesh. A number of local cultivars like Badrakali chuvala, Chakkaravalli, Anakomban and Kottaram chuvala are grown mainly in Kerala. Central Tuber Crops Research Institute, in Thiruvanthapuram has developed several good varieties like Varsha, Sree Nandini, Sree Vardhini, Sree Ratna, Shree Bhadra, Sree Arun, Sree Varun. IARI, New Delhi released Pusa Safed and Pusa Sunheri. ANGRAU, Hyderabad developed Cross-4, Rajendra Shakarkand-5 and Kalmegh. Tamil Nadu Agricultural University, Coimbatore released Co-1 and Co-2.

Uses

The main use of sweet potato is for human consumption where it is usually eaten after boiling, baking and frying. It is also an important source of starch, glucose, pectin sugar syrup and industrial alcohol.

Nutritional Components of Sweet Potato

Constituents	Quantity (per 100 g)	Constituents	Quantity (per 100 g)
Moisture	70 g	Sulphur	26 mg
Carbohydrates	27 g	Potassium	373 mg
Proteins	1.5-2.0 g	Sodium	13 mg
Fats	0.2 g	Chlorine	85 mg
Fibers	1.0 g	Carotene	18 mg
Energy	120 Kcal	Vitamin C	24 mg
Calcium	46 mg	Thiamine	0.08 mg
Iron	0.8 mg	Niacin	0.70 mg
Magnesium	24 mg	Riboflavin	0.04 mg
Phosphorus	49 mg		

Gopalan et al., (1985)

Phytochemicals

Sweet potato has played an important role as an energy and a phytochemical source in human nutrition and animal feeding. Ethnopharmacological data show that sweet potato leaves have been effectively used in herbal medicine to treat inflammatory and/or infectious oral diseases in Brazil. The sweet potato could be considered as an excellent source of natural health-promoting compounds, such as beta-carotene and anthocyanins, for the functional food market (Bovell, 2007). The phytochemical screening showed presence of triterpenes/steroids, alkaloids, anthraquinones, coumarins, flavonoids, saponins, tannins, and phenolic acids. The antioxidant capacities in sweet potato is reported to be due to the high levels of phenols, flavonoids, β-carotene, anthocyanins, and caffeoylquinic acid derivatives (Dini et al., 2009; Carvalho et al., 2010; Rumbaoa et al., 2009).

Sweet potato leaves contain at least 15 anthocyanin and 6 polyphenolic compounds. These biologically active compounds possess multifaceted action, including antioxidation, antimutagenicity, anti-inflammation and anticarcinogenesis. Sweet-potato leaves contain more total polyphenols than any other commercial vegetables, including sweet potato roots and potato tubers.

One of the more intriguing nutrient groups provided by sweet potatoes—yet one of the least studied from a health standpoint—are the resin glycosides. These nutrients are sugar-related and starch-related molecules that are unusual in their arrangement of carbohydrate-related components, and also in some non-carbohydrate molecules. In sweet potatoes, researchers have long known of one group of resin glycosides called 'batatins' (including batatin I and batatin II). But only recently have researchers discovered a related group of glycosides in sweet potato called 'batatosides' (including batatoside III, batatoside IV, and batatoside V). In lab studies, most of these sweet potato glycosides have shown to possess antibacterial and antifungal properties.

Sweet potatoes have antioxidants, anti-inflammatory nutrients, and blood sugar-regulating nutrients. Sweet potatoes contain a wealth of orange-hued carotenoid pigments. In some studies, sweet potatoes have been shown as a better source of bioavailable beta-carotene than green leafy vegetables. Particularly in purple-fleshed sweet potato, anthocyanin pigments are abundant. Cyanidins and peonidins are concentrated in the starchy core of part of purple-fleshed sweet potatoes, and these antioxidant nutrients may be even more concentrated in the flesh than in the skin. Extracts from the highly pigmented and colorful purple-fleshed and purple-skinned sweet potatoes have been shown in research studies to increase the activity of two key antioxidant enzymes—copper/zinc superoxide dismutase (Cu/Zn-SOD) and catalase (CAT).

Health Benefits of Sweet Potato

Recent research has shown that sweet potato cyanidins and peonidins and other color-related phytonutrients may be able to lower the potential health risk posed by heavy metals and oxygen radicals. It is also reported to cure digestive tract problems like irritable bowel syndrome or ulcerative colitis.

Anti-Inflammatory: In the case of inflammation, in animal studies, activation of nuclear factor-kappa B (NF-Î°B); activation of inducible nitric oxide synthase (iNOS); and cyclooxygenase-2 (COX-2); and formation of malondialdehyde (MDA) have all been shown to get reduced following consumption of either sweet potato or its color-containing extracts. Similarly, studies showed reduced inflammation in brain tissue and nerve tissue throughout the body following sweet potato consumption.

Blood Sugar Regulator: The blood sugar regulating effect of sweet potato is believed to be brought out by modifying adiponectin. Adiponectin is a protein hormone produced by our fat cells, and it serves as an important modifier of insulin metabolism.

Estrogen hormone Regulator: Sweet potatoes have been labeled as natural alternative to estrogen therapy or a natural dehydroepiandrosterone because laboratory studies have shown that they contain a chemical called 'diosgenin', which can be changed into different steroids, such as estrogen. Sweet potatoes are used as estrogen replacement therapy, as well as for managing menstrual problems and osteoporosis.

Other uses include reducing diverticulosis, which is an ailment of the intestines; gall bladder complaints; and arthritis. Available scientific information does not support the clinical efficacy of sweet potatoes for any other health condition.

Therapeutic properties of Sweet Potato

Several researchers reported that sweet-potato leaves have anti-diabetic compounds that reduce the blood glucose content significantly in model rats. The water extracted from the leaves suppressed effectively the growth of other food-poisoning bacteria such as *Staphylococcus aureus* and *Bacillus cereus* as well as pathogenic *E. coli* (Islam, 1977).

Consumption of Purple Sweet Potato leaves modulate various immune functions including increased proliferation responsiveness of PBMC (peripheral blood mononuclear cells), secretion of cytokines IL-2 and IL-4, and the lytic activity of NK cells (Chen *et al.*, 2005).

The study by Runnie *et al.*, (2004) demonstrated that many edible plants common in Asian diets like sweet potato (white and purple) possess potential health benefits, affording protection at the vascular endothelium level. Ludvick *et al.*, (2003) described the underlying mechanism responsible for the improvement in metabolic control following administration of sweet potato in type 2 diabetes subjects. They found that short-term treatment with 4 g/d of the sweet potato-based nutraceutical consistently improved metabolic control in type 2 diabetic patients by decreasing insulin resistance without affecting the body weight, glucose effectiveness, or insulin dynamics.

Purple sweet potato leaves (PSPL), which are easily grown in tropical areas, have the highest polyphenolic content, in particular flavonoids, of all the commonly grown vegetables, and exhibit free radical scavenging ability (Chu *et al.*, 2000). Storage proteins in sweet potato called 'sporamins' also have important antioxidant properties.

Park *et al.*, (2010) first showed the biological functions of Purple Sweet Potato (PSP) extract in treating hyperlipidemic and hyperglycemic disorders in rats. When streptozotocin-induced diabetic rats were treated wih daily oral feeding of powdered *Ipomoea batatas* (5 g/kg body weight/day), for 2 months it significantly decreased the level of fasting plasma glucose and hemoglobin A1c levels, and restored body weight loss during diabetes (Niwa *et al.*, 2011). Jiang *et al.*, (2011) found that diabetic symptoms were ameliorated after rats were fed with Purple Sweet Potato Flavonoids.

The fasting blood glucose (FBG), GSP, TC, TG, LDL-C were decreased and serum HDL-C levels were increased in high, medium dose PSPF groups. Oki *et al.*, (2011) reported significant decrease in plasma glucose levels by administering white skinned sweet potato- arabino galactanprotein (WSSP-AGP). Kusano *et al.*, (2001) isolated and studied the antidiabetic component in white skinned sweet potato responsible for its functional property and reported that the active component is presumed to be an acidic glycoprotein because it contained protein and sugar and was adsorbed onto the QA column at pH 7.0.

References

Bovell-Benjamin, AC. 2007. Sweet potato: a review of its past, present, and future role in human nutrition. Adv Food Nutr Res., (2007), 52:1-59.

Carvalho, I.S., Cavaco, T., Carvalho, L.M., and Duque P. 2010. Effect of photoperiod on flavonoid pathway activity in sweet potato (*Ipomoea batatas* (L.) Lam.) leaves. Food Chem., (2010),118:384–90.

Chen, C.M., Li, S.C., Lin, Y.L., Hsu, C.Y., Shieh, M.J., and Liu, J.F. 2005.Consumption of purple sweet potato leaves modulates human immune response: T-lymphocyte functions, lytic activity of natural killer cell and antibody production. World J Gastroenterol., (2005), 11(37):5777-81.

Chu, Y.H., Chang, C.L., and Hsu, H.F. 2000. Flavonoid content of several vegetables and their antioxidant activity. J Sci Food Agr., (2000), 80: 561-566.

Dini, I., Tenore, G.C., and Dini, A. 2009. Saponins in *Ipomoea batatas* tubers: Isolation, characterization, quantification and antioxidant properties. Food Chem., (2009), 113:411–9.

http://www.whfoods.com/genpage.php?tname=foodspice&dbid=64.

Jiang, H.F., Li, X.R., and Tang, C. 2011. Effect of purple sweet potato flavonoids on metabolism of glucose and lipids in diabetic rats. Zhejiang Da Xue Xue Bao Yi Xue Ban. (2011), 40(4):374-9.

Kusano, S., Abe, H., and Tamura, H. 2001. Isolation of antidiabetic components from white-skinned sweet potato (*Ipomoea batatas* L.). Biosci Biotechnol Biochem., (2001), 65(1):109-14.

Ludvik, B., Waldhäusl, W., Prager, R., Kautzky-Willer, A., and Pacini, G. 2003. Mode of action of ipomoea batatas (Caiapo) in type 2 diabetic patients. Metabolism, (2003), 52(7):875-80.

Niwa, A., Tajiri, T., and Higashino, H. 2011. *Ipomoea batatas* and *Agarics blazei* ameliorate diabetic disorders with therapeutic antioxidant potential in streptozotocin-induced diabetic rats. J Clin Biochem Nutr., (2011), 48(3):194-202.

Oki N., Nonaka, S., and Ozaki, S. 2011.The effects of an arabinogalactan-protein from the white-skinned sweet potato (*Ipomoea batatas* L.) on blood glucose in spontaneous diabetic mice. Biosci Biotechnol Biochem., (2011), 75(3):596-8.

Park, K.H., Kim, J.R., Lee, J.S., Lee, H., and Cho, K.H. 2010. Ethanol and water extract of purple sweet potato exhibits anti-atherosclerotic activity and inhibits protein glycation. J Med Food., (2010), 13(1):91-8.

Rumbaoa, R.G., Cornago, D.F., and Geronimo, I.M. 2009. Phenolic content and antioxidant capacity of Philippine sweet potato (*Ipomoea batatas*) varieties. Food Chem., (2009),113:1133–8.

Runnie, I., Salleh, M.N., Mohamed, S., Head, R.J., and Abeywardena, M.Y. 2004. Vasorelaxation induced by common edible tropical plant extracts in isolated rat aorta and mesenteric vascular bed. J Ethnopharmacol., (2004), 92(2-3):311-6.

Islam, S.1977. Medicinal and nutritional qualities of sweetpotato tops and leaves. Plant Science University of Arkansas at Pine Bluff, United States Department of Agriculture, and County Governments Cooperating. FSA6135.

6.3. Taro (*Colocasia esculenta* Schott)

Taro is a large perennial herbaceous plant growing up to 5-6 feet and belongs to the family Araceae. It grows best in marshy, wet soil and warm humid climates. Inside its flesh has white to cream yellow color but may have different colors depending on cultivar types. Its delicious flesh is crisp in consistency and water chestnut like nutty flavor.

Taro (Colocasia) originated in South East Asia including India and Malaysia. There are two types *C. esculenta* var. *esculenta* and *C. esculanta* var. *antiquorum*. The var. *esculenta* consists of wild types like Naga kaju of Nagaland, Khasi Bhunga of Meghalaya, Kovur of Andhra Pradesh, local cultivars like Panch Mukhi, Koni Kachu and White Gauriya. *C.esculenta* var. *antiquorum* includes both wild and cultivated forms. There can be variations in the nutrients content of each ecotype as it is controlled by various factors.

In Taro there are diverse colours, shapes, and sizes. The edible part tuber is an underground stem. Due to acridity, tubers are used as vegetable after cooking only. In India it is used as a vegetable, while in Africa, it is used for making a product called '**fufu**'. Fufu, is commonly made by pounding starchy food crops such as yam, cassava, plantain and others with hot water. It's eaten throughout the West African region and in several Caribbean countries including Haiti, Jamaica, Cuba and Puerto Rico. In Hawai and Polynesia it is used for making a fermented acidic product called '**Poi**'. It is recommended for gastric patients and also for preparing baby food.

Yautia (*Xanthosoma* species), also known as *tannia, malanga etc*, is similar to taro but smaller and has somewhat elongated, bumpy corms grown widely in East Asia, Caribbean and South American regions.

Eddoe (*Colocasia esculenta antiquorum*) is also smaller corm with irregular surface (like ginger) grown widely in India, China, and Japan as well as in some Caribbean countries. It is known as '*arbi*' or 'arvi'in India.

Cultivars/Varieties: Taro cultivars are named after the shape of the corms as Panch Mukhi, Sahasra Mukhi and Tamarakannan. The variety 'Kannanchembu' is famous for its excellent quality. CTCRI, Thiruvananthapuram also released a few varieties like Sree Rashmi, Sree Pallavi, and a F_1 hybrid called 'Sree Kiran'.

Uses

Taro's primary use is the consumption of its edible corm and leaves. In its raw form, the plant is toxic due to the presence of calcium oxalate and the presence of needle-shaped raphides in the plant cells. However, the toxin can be destroyed, and the tuber rendered palatable by cooking, or by steeping in cold water overnight. Corms of the small round variety are peeled and boiled, sold either frozen, bagged in its own liquids, or canned. The leaves are rich in vitamins and minerals. It is a rich source of beta carotene and calcium. Taro is extremely important to the Hawaiians who associated it with their Gods and their story of creation, and even used it for medicinal purposes.

Nutritional Components of Taro

Constituents	Quantity (per 100 g)	Constituents	Quantity (per 100 g)
Moisture	82.7 g	Iron	10 mg
Carbohydrates	6.8 g	Calcium	40 mg
Proteins	3.9 g	Phosphorus	140 mg
Fats	1.5 g	Carotene	24 mg
Energy	97 Kcal	Thiamine	0.09 mg
Calcium	0.26 mg	Niacin	0.4 mg

Gopalan *et al.*, (1985)

Taro corms are rich in carbohydrates, and free from gluten. They have high quality phyto-nutrition profile comprising of dietary fiber, and antioxidants in addition to moderate proportions of minerals, and vitamins.

Phytochemicals

Forty-one phenolic metabolites (11 hydroxycinnamic acid derivatives and 30 glycosylated flavonoids) were identified by high-performance liquid chromatography-diode array detection-electrospray ionization/mass spectrometry (HPLC-DAD-ESI/MS(n)) in the leaves of two *C. esculenta* varieties cultivated in Azores Islands. "Giant white" and "red" varieties (local denomination) contain, respectively, 14 and 21% of phenolic acids, 37 and 28% of flavones mono-C-glycosides, 42 and 43% of flavones di-C-glycosides, 3 and 4% of flavones mono-C-(O-glycosyl) glycosides, and both of them have ioavai.. 2% of flavones di-C-(O-glycosyl) glycosides and 2% of flavones-O-glycosides. Luteolin-6-C-hexoside was the compound identified to be in higher amounts in both varieties (Ferreres *et al.*, 2012). Hydro-alcoholic extract of *Colocasia esculenta* showed presence of flavonoids, β-sitosterol, and steroids (Kalariya *et al.*, 2010).

Health benefits of Taro

- Taro tubers and leaves helps in reducing the sugar levels in diabetics and also reduce the level of cholesterol in hyperlipidemia patients.
- Taro leaves as well as yellow-fleshed roots have significant levels of phenolic flavonoid pigments (antioxidants) such as *β-carotenes, and cryptoxanthin* along with vitamin A. Consumption of natural foods rich in flavonoids helps to protect from lung and oral cavity cancers.
- It also contains good levels of some of valuable B-complex group of vitamins such as pyridoxine (vitamin B-6), folates, riboflavin, pantothenic acid, and thiamin.
- Taro is rich in potassium which is an important component of cell and body fluids that help regulate heart rate and blood pressure.
- 'Poi' is a fermented common food form of taro consumed in Hawaii. Fermentation proceeds without inoculated starter cultures and usually takes about two to three days to reach the "sour" stage. Studies several decades ago suggested the use of poi in treating certain medical conditions, especially infant food allergies and failure-to-thrive in infants. Another potential medicinal use of poi is as a probiotic because it contains the predominant lactic acid bacteria (*L. lactis*).

Therapeutic Properties of Taro

Ethanol extract of *Colocasia esculenta* (taro) showed the highest inhibition of human lanosterol synthase (hOSC) (55% inhibition at 300m g/ml), the compounds responsible for suppression of cholesterol biosynthesis. Further purification and fractionation of the extract showed three monogalactosyldiacyl-glycerols and five digalactosyldiacyl-glycerols as active compounds that showed 28 to 67% inhibitory activities at the concentration 300m g/ml (Yuichi *et al.*, 2005). Taro root possess amylopectin and amylase, which are more complex carbohydrates that contribute to slow digestion, and regulates blood glucose levels.

Kundu *et al.*, (2012) first reported compound(s) derived from taro that potently and specifically inhibits tumor metastasis. They have shown that a water-soluble extract of taro (TE) potently inhibits lung-colonizing ability and spontaneous metastasis from mammary gland-implanted tumors, in a murine model of highly metastatic estrogen receptor, progesterone receptor and Her-2/neu-negative breast cancer. TE modestly inhibited the proliferation of some, but not all, breast and prostate cancer cell lines. Tumor cell migration was completely blocked by TE. TE treatment also inhibited prostaglandin E2 (PGE2) synthesis and downregulated cyclooxygenase 1 and 2 mRNA expression, indicating its anti-inflammatory effect.

In a study conducted to identify traditional remedies used for treating diabetic ailments in North Eastern India, among the 21 plant species reported, one of them was *Colocasia esculanta* (Taro) which was very important in the primary health care of the people living in rural Dhemaji district of Assam, north-east India (Tarak *et al.*, 2011).

Adult T-cell leukaemia (ATL) is caused by human T-cell leukaemia virus type I (HTLV-I) infection and is resistant to conventional chemotherapy. Kai *et al.*, (2011) evaluated the inhibitory effects of agricultural plants on the proliferation of seven ATL-related human leukaemia cells, using three ATL cell lines (ED, Su9T01 and S1T), two human T-cell lines transformed by HTLV-I infection (HUT-102 and MT-2) and two HTLV-I-negative human T-cell acute lymphoblastic leukaemia cell lines (Jurkat and MOLT-4). Their results showed that extracts from edible parts of *Ipomea batatas* LAM. (sweet potato), edible parts of *Colocasia esculenta* (L.) Schott (taro) and skin of taro showed markedly greater inhibitory effects on Su9T01 than genistein which was used as a positive control.

In order to examine how chemically diverse fibres differ in their hypolipidaemic activity, mucilages of varying chemical composition isolated from three different sources, galactomannan isolated from fenugreek seeds, a glucomannan from *Dioscorea esculenta* tubers and an arabinogalactan from *Colocasia esculenta* tubers were administered to experimental animals and the metabolism of lipids and lipoproteins was studied (Boban *et al.*, 2006). Among the different mucilages, mannose-rich glucomannan from Taro (*Colocasia esculenta*) showed the most effect followed by galactomannan, and mannose-free arabinogalactan.

Hawaiians tend to have lower incidence rates of colorectal cancer and it was hypothesized that this may be due to ethnic differences in diet, specifically, their consumption of poi, a starchy paste made from the taro (*Colocasia esulenta* L.) plant corm. A study conducted by Brown *et al.*, (2004) indicated that although numerous factors can contribute to the risk of colon cancer, perhaps 'poi' consumption may contribute to the lower colon cancer rates among Hawaiians by two distinct mechanisms. First, by inducing apoptosis within colon cancer cells; second, by non-specifically activating lymphocytes, which in turn can lyse cancerous cells.

References

Brown, A.B., and Valiere, A.M.S. 2004. The Medicinal Uses of Poi. *Nutr Clin Care*, (2004), 7(2): 69–74.

Boban, P.T., Nambisan, B, and Sudhakaran, P.R. 2006. Hypolipidaemic effect of chemically different mucilages in rats: a comparative study. Br J Nutr., (2006), 96(6):1021-9.

Ferreres, F., Gonçalves, R.F., Gil-Izquierdo, A., Valentão, P., Silva, A.M., Silva, J.B., Santos, D., and Andrade, P.B. 2012. Further knowledge on the phenolic profile of *Colocasia esculenta* (L.) Shott. J Agric Food Chem., (2012), 60(28):7005-15.

http://www.nutrition-and-you.com/taro.html

Kai, H., Akamatsu, E., Torii, E., Kodama, H., Yukizaki, C., Sakakibara, Y., Suiko, M., Morishita, K., Kataoka, H., and Matsuno, K. 2011. Inhibition of proliferation by agricultural plant extracts in seven human adult T-cell leukaemia (ATL)-related cell lines. J Nat Med., (2011), 65(3-4):651-5.

Kalariya, M., Parmar, S., and Sheth, N. 2010. Neuropharmacological activity of hydroalcoholic extract of leaves of *Colocasia esculenta*. Pharm Biol., (2010), 48(11):1207-12.

Kundu, N., Campbell, P., Hampton, B., Lin, C.Y., Ma, X., Ambulos, N., Zhao, X.F., Goloubeva, O., Holt, D., and Fulton, A.M. 2012. Antimetastatic activity isolated from *Colocasia esculenta* (taro). Anticancer Drugs., (2012), 23(2):200-11.

Tarak, D., Namsa, N.D., Tangjang, S., Arya, S.C., Rajbonshi, B., Samal, P.K., and Mandal M. 2011. An inventory of the ethnobotanicals used as anti-diabetic by a rural community of Dhemaji district of Assam, Northeast India. J Ethnopharmacol., (2011), 138(2):345-50.

Sakano, Y., Matsuga, M., Tanaka, R., Suganuma, H., Inakuma, T., Toyoda, M., Gods, Y., Shibuya, M., and Ebizuka, Y. 2005. Inhibition of human lanosterol synthase by the constituents of *Colocasia esculenta* (Taro). Biol. Pharm. Bull., 28(2) 299—304.

6.4. Cassava (*Manihot esculenta* L.) or Tapioca

Cassava (*Manihot esculenta* L.) belongs to the family Euphorbiaceae. It is native to North Easten Brazil and is believed to have been introduced to India by Portuguese. It is an underground tuber and is rich in starch and is mainly consumed after cooking.

Tapioca was used by the first inhabitants of the West Indies as a staple food from which they made main dishes such as pepper pot and also used it to make alcohol. It was also used to clean their teeth and till today it is used as a base in toothpaste. Currently it is still a very popular food in the islands, used as a provision cooked with meats or fish and in desserts such as cassava pone.

In various Asian countries such as Indonesia, China, India, Pakistan, Bangladesh, Myanmar, Philippines, Malaysia, and Taiwan, tapioca pearls are widely used and are known as sagudana, sabudana (pearl sago) or "sabba akki". The pearls (sagudana or shabudana/sabudana) are used to make snacks or for pudding/payasa. In India, cassava is cultivated in southern peninsular region, particularly Kerala, Tamil Nadu and Andhra Pradesh contributing to 93% of area and 98% of production.

Cultivars/Varieties: Most of the improved varieties are released from CTCRI, Thiruvananthapuram are H-97, H-165, H-226, Sree Sahya, Sree Vaisakham, Sree Prakash, Sree Harsha, Sree Jaya, Sree Vijaya, Sree Rekha and Sree Prabha. Kerala Agricultural University released three varieties viz. Nidhi, Kalpaka and Vellayani Hraswa. Co-1, Co-2 and Co-3 are the varieties from Tamil Nadu Agricultural University, Coimbatore.

Uses

Tapioca is widely consumed in the state of Kerala. It is either boiled or cooked with spices. Tapioca with fish curry (especially Sardine) is a delicacy in Kerala. Thinly sliced tapioca wafers, similar to potato chips, are popular too. Cassava, often referred to as tapioca in English, is called Kappa Kizhangu or Poola (in northern Kerala) or Maracheeni or Cheeni or Kolli or Mathock in Malayalm. Tapioca is regarded as a staple food of the common man in Kerala.

It is gluten and protein free and has some nutritional value. Tapioca is a carbohydrate rich crop, mostly comprising starch. The nutritional value of tapioca is usually determined by what recipe it is used in. Cassava tubers contain calories, protein, fat, carbohydrate, calcium, phosphorus, iron, vitamin B and C, and starch. The leaves contain vitamins A, B1 and C, calcium, calories, protein, fat, carbohydrate, and iron, while the bark, contains tannin, peroxidase enzymes, glycosides, and calcium oxalate.

Nutritional Components of Tapioca (Cassava)

Constituents	Quantity (per 100 g)	Constituents	Quantity (per 100 g)
Moisture	59.40 g	Iron	0.90 mg
Carbohydrates	38.10	Phosphorus	40 mg
Proteins	0.70 g	Carotene	0.00
Fats	0.20 g	Thiamine	0.05 mg

Constituents	Quantity (per 100 g)	Constituents	Quantity (per 100 g)
Energy	157 Kcal	Riboflavin	0.10 mg
Calcium	50 mg	Niacin	0.03 mg

Gopalan *et al.*, (1985)

Phytochemicals

Besides the above nutritional constituents, several othe phytochemicals were also reported to be present in cassava. Yi *et al.*, (2011) isolated ten phenolic compounds: Coniferaldehyde (1), isovanillin (2), 6-deoxyjacareubin (3), scopoletin (4), syringaldehyde (5), pinoresinol (6), p-coumaric acid (7), ficusol (8), balanophonin (9) and ethamivan (10) from cassava which was reported to be responsible for its antioxidant activity. The first isolation and identification of galactosyl diacylglycerides from fresh cassava roots was reported by Bayoumi *et al.*, (2010), as beta-carotene, linamarin, and beta-sitosterol glucopyranoside. Hydroxy-coumarin scopoletin and its glucoside scopolin were also identified from cassava roots.

Health Benefits of Cassava Based on Traditional Knowledge and Folklore Practices

- Cassava has nearly twice the calories than potatoes, perhaps highest for any tropical starch rich tubers and roots. The majority of carbohydrates is in the form of starch, besides some sugars. 100 g root provides 160 calories, which mainly comes from sucrose forming the bulk of the sugars in tubers, accounting for more than 69% of the total sugars. Complex sugar amylose is another major carbohydrate source (16-17%).

- Cassava is very low in fats and protein than in cereals and pulses. Nonetheless, it has more protein than that of other tropical food sources like yam, potato, plantain, *etc.*

- Similar to other roots and tubers, cassava too is free from gluten. Gluten-free starch is used in special food preparations for celiac disease patients.

- Young tender cassava leaves and root are good source of dietary proteins and vitamin K. Vitamin-K has important role in bone mass building by promoting osteotrophic activity in the bones.

- It also has established role in the treatment of Alzheimer's disease patients by limiting neuronal damage in the brain.

- Cassava is a moderate source of some of valuable B-complex group of vitamins such as folates, thiamin, pyridoxine (vitamin B-6), riboflavin, and pantothenic acid.

- The root is the chief source of some important minerals like zinc, magnesium, copper, iron, and manganese for many inhabitants in the tropical belts. In addition, it has adequate amounts of potassium (271 mg per 100g or 6% of RDA). Potassium is an important component of cell and body fluids that help regulate heart rate and blood pressure.

- Cassava is a tuber that contains a lot of insoluble fiber or rouphages that is not soluble in water. Fibers of this type facilitate easy defecation, as it absorbs waster makes the stools bulky. It also absorbs and remove toxins from the intestine and maintains healthy digestive system.

- Traditional way to relieve pain due to rheumatism, is to boil 100 grams of cassava stems, one stem of lemon grass, and 15 grams ginger with 1000 ml of water, and then, filter and drink the water, twice daily.
- Crushed fresh cassava stem when placed on woulds or scratches and bandaged, will help in faster healing.
- For controlling diarrhea, boil seven pieces of cassava leaves with 800 ml of water and concentrate 400 ml by continuous boiling. Then filter and drink the extract 200 ml twice daily.
- For deworming, boil 60 grams of cassava bark and 30 grams of leaves with 600 ml of water and boil until it reaches 300 ml. Then filter and drink the water before going to bed.
- Beri-beri, can be overcome by consuming cassava leaves as salad.
- In chinese medicine, cassava is used to improve stamina. 100 grams of cassava, 25 grams red ginger (*Kaempferia galanga)*, and five red dates (deseeded), blended with enough water is added with honey and consumed as a rejuvenating drink.

Antinutritional Factors

Raw cassava contains cyanogenic glucosides, particularly linamarin and lotaustralin, which are converted to prussic acid (hydrogen cyanide, HCN) when they come in contact with linamarase, an enzyme that is released when the cells of cassava roots are ruptured. Good processing and cooking methods reduce the cyanide levels, and acute cyanide toxicity rarely occurs. Chronic cyanide toxicity occurs when cassava consumption is very high (7 kg or more of fresh roots per day over long periods), especially where iodine and/or protein consumption is very low.

Therapeutic Properties of Cassava

Rahmat *et al*., (2003) conducted a cytotoxic study, where the ethanolic extract of cassava was tested on several cell lines i.e. breast cancer (MDA-MB-231 and MCF-7), colon cancer (Caco-2), liver cancer (HepG2) and normal liver (Chang liver). The findings suggested that the tapioca shoot ethanolic extract was able to inhibit the viability of MCF-7 cell line with the IC50 value of 52.49 µg/ml. They concluded that tapioca shoots have potential as an anticancer agent against certain breast tumours.

The studies of Boby and Indira (2003) suggested that the fiber and antioxidant vitamins present in the cassava may be playing a protective role against the alcohol induced oxidative stress. However, one study by Yessoufou *et al*., (2006) has shown that cassava aggravates diabetes. As cassava has certain anti-nutritional factors, it is advised to consume with proper cooking or processing to get its potential health benefits.

References

Ayodeji, O., and Fasuyi. 2005. Nutritional evaluation of cassava (Manihot esculenta, Crantz) leaf protein concentrates (CLPC) as alternative protein sources in rat assay. Pakistan Journal of Nutrition, 4 (1): 50-56.

Bayoumi, S.A, Rowan, M.G., Beeching, J.R., and Blagbrough, I.S. 2010. Constituents and secondary metabolite natural products in fresh and deteriorated cassava roots. Phytochemistry, (2010), 71(5-6):598-604.

Boby, R.G., and Indira, M. 2003. The impact of cyanoglycoside rich fraction isolated from Cassava (Manihot esculenta) on alcohol induced oxidative stress. Toxicon., (2003), 42(4):367-72.

Gopalan, C., Sastri, B.V.R., and Balasubramanian., S.C. 1985. Nutritive Value of Indian Foods. National Institue of Nutrition, ICMR, Hyderabad, India.

http://benefits-of-cassava.weebly.com/

http://en.wikipedia.org/wiki/Tapioca

http://traditionalmedicinecenter.blogspot.in/2010/03/cassava-for-health-benefits-cassava.html

http://www.food-info.net/uk/products/rt/cassava.htm

Rahmat, A., Kumar, V., Fong, L.M., Endrini, S., and Sani, H.A. 2003. Determination of total antioxidant activity in three types of local vegetables shoots and the cytotoxic effect of their ethanolic extracts against different cancer cell lines. Asia Pac J Clin Nutr., (2003), 12(3):292-5.

Yessoufou, A., Ategbo, J.M., Girard, A., Prost, J., Dramane, K.L., Moutairou, K., Hichami, A., Khan, N.A. 2006. Cassava-enriched diet is not diabetogenic rather it aggravates diabetes in rats. Fundam Clin Pharmacol., (2006), 20(6):579-86.

Yi, B., Hu, L., Mei, W., Zhou, K., Wang, H., Luo, Y., Wei, X., and Dai, H. 2011. Antioxidant phenolic compounds of cassava (Manihot esculenta) from Hainan, (2011), 16(12):10157-67.

6.5. Elephant Foot Yam

(Amorphophallus campanulatus Blume) Syn. *Amorphophallus paeoniifolius*
Dennst.) Nicolson, *Amorphophallus konjac* (K.Koch)

Amorphophallus or Elephant – foot-yam otherwise c alled Suran or Jimmikand is a popular vegetable in certain areas of the tropical and subtropical regions. It belongs to the family Araceae. Amorphophallus is cheap sources of carbohydrates and rich in minerals and vitamin A and B. Though poor in protein per se, it exhibits a high degree of variation in its protein content depending on cultivars. The corm is used as vegetable and is good as secondary food during the period of scarcity. It is also used for preparing curries and pickles. There are two cultivars, namely Santragachi and Kovvur which are very popular in eastern and southern parts of India.

Nutritive components of Elephant Foot Yam

Constituents	Quantity (per 100 g)	Constituents	Quantity (per 100 g)
Moisture	78.7 g	Iron	0.60 mg
Carbohydrates	18.4 g	Phosphorus	34 mg
Proteins	1.2 g	Carotene	260 mg
Fats	0.10 g	Thiamine	0.06 mg
Energy	79 Kcal	Riboflavin	0.07 mg
Calcium	50 mg	Niacin	0.70 mg

Gopalan *et al.*, (1985)

Uses

Elephant foot yam is basically a crop of Southeast Asian origin. It grows in wild form in the Philippines, Malaysia, Indonesia, and other Southeast Asian countries. In India it is grown mostly in West Bengal, Kerala, Andhra Pradesh, Maharashtra and Odisha. Elephant foot yam is used traditionally for the treatment of tumours, piles, abdominal pain, and enlargement of spleen, asthma and also in rheumatism. Most of the studies showed that in Siddha medicine *Amorphophallus campanulatus* is used in the treatment of piles. Amorphophallus roots has many medicinal uses and it is a good source of phyto-estrogen and are effective alternate or complementary hormone therapy for symptoms associated with menopause and chronic degenerative disease in women.

Cultivars/Varieties: There are two improved varieties, Sree Padma from CTCRI, Thiruvananthapuram and Gajendra from APAU, Hyderabad.

Phytochemicals

The photochemical screening of Amorphophullus corm revealed the presence of carbohydrates, alkaloids. Tri-terpenoids, coumarins, steroids, tannins, saponins., flavones, flavanones, quinines, anthraquinones anthocyanins, phenols. Proteins and aromatic amino acids (Firdose and Alam, 2011). The corm contains diastatic enzyme amylase, betulinic acid, lupeol, stigmasterol, beta-sitosterol and its palmitate.

Among the carbohydrates, glucose, galactose, rhamnose, triacontane, niacin, carotene and xylose were the prominent components (Khare, 2007; Sahu *et al.*, 2009). A water-soluble polysaccharide isolated from the aqueous extract of the corm of *Amorphophallus campanulatus* was found to contain D-galactose, D-glucose, 4-O-acyl-D-methyl galacturonate, and l-arabinose in a molar ratio 2:1:1:1 (Das *et al.*, 2009).

Health Benefits of Amorphophallus

- Elephant foot yam has many health benefits and therapeutic uses. The tubers are anodyne, anti-inflammatory, antihaemorrhoidal, haemostatic, expectorant, carminative, digestive, appetizer, stomachic, anthelmintic, liver tonic, aphrodisiac, emmenagogue, rejuvenating and tonic. They are traditionally used in arthralgia, elephantiasis, tumors, inflammations, hemorrhoids, hemorrhages, vomiting, cough, bronchitis, asthma, anorexia, dyspepsia, flatulence, colic, constipation, helminthiasis hepatopathy, splenopathy, amenorrhea, dysmenorrhoea, seminal weakness, fatigue, anemia and general debility (Nair, 1993).

- The tuber is reported to have anti-protease activity due to which it acts as anti-inflammatory (Prathiba *et al.*, 1995), analgesic activity due to which it gives relief from pain (Shipli *et al.*, 2005), and cytotoxic activity resulting in its anticancerous property (Angayarkanni *et al.*, 2007).

- The tuber is widely used in many Ayurvedic preparations and prescribed in liver diseases, bronchitis, asthma, abdominal pain, emesis, dysentery, enlargement of spleen, piles, elephantiasis, blood and rheumatic swellings (Ansil *et al.*, 2012).

- *Amorphophallus konjac* has an effect on diabetes as well. Its ability to move through the digestive tract very slowly also slows down carbohydrate absorption. This slowed absorption will keep the blood sugar at a moderate level.

- *Amorphophallus konjac* also acts to trap food and waste residues as it moves through the digestive system. This internal scrubbing opens the pores along the digestive tract, which in turn increases or regulates absorption.

- Its presence can also block substances that are easily reabsorbed such as excess bile acids.

- The digestive regulatory quality of *Amorphophallus konjac* makes it a natural, alternative agent in the treatment and regulation of diabetes and high cholesterol.

Therapeutic properties of Elephant foot yam

Antioxidant & Hepatoprotective: Elephant foot yam is reported to have antioxidant and hepatoprotective properties. Singh *et al.*, (2011) has demonstrated that the methanolic extract of *Amorphophallus campanulatus* has hepatoprotective activity against acetaminophen induced hepatotoxicity in rats and it may be due to their antioxidant property. Further, the studies of Ansil *et al.*, (2012) also indicated that the antioxidant and hepatoprotective effect of the *Amorphophallus campanulatus* tuber extract might be attributed to the presence of the identified class of chemicals in single or in combination. The possible mechanism behind the antioxidant and hepatoprotective property of ACME may be associated with stimulation of antioxidant defense mechanism against the free radicals generated by TAA or by the inhibition of cytochrome-P450 enzyme system responsible for the generation of the toxic free radicals from these chemicals. The results of the study of Sahu *et al.*, (2009) showed that the extract of *Amorphophallus campanulatus* tuber had good antioxidant and free radical scavenging activity. The aqueous extract showed more significant antioxidant activity as compared to methanol extract.

The study also revealed that *Amorphophallus campanulatus* tuber could be resourcefully used for the development of a phytomedicine against oxidative stress and/or liver ailments (Ansil *et al.*, 2012). Jain *et al.*, (2009) found potent hepatoprotective action of ethanolic and aqueous extracts of *Amorphophallus campanulatus* (Roxb.) tubers against carbon tetrachloride induced hepatic damage in rats.

Anti-microbial property: Khan *et al.*, (2008) conducted studies to see the antibacterial and antifungal activities of amblyone, a tri-terpenoid isolated from *Amorphophallus campanutus* tubers and reported that it was very effective against the gram-positive bacteria like *Bacillus subtilis, Bacillus megatarium, Staphylococcus aureus and Streptococcus pyogenes*. Amblyone showed weak antifungal activity against many fungi. The cytotoxic effect of Amblyone was proved and found to be as moderately effective as gallic acid or vincristine sulphate.

Analgesic & Anti-inflammatory: The methanol extract of *Amorphophallus campanulatus* tuber, given orally at the doses of 250 and 500 mg/kg, was also found to show significant analgesic activity in mice (Shilpi *et al.*, 2005).

Anti-Diabetic: The antidiabetic effect of *Amorphophallus konjac* glucomannan with different molecular chains on experimental diabetes mice was studied by Li *et al.*, (2004) which showed that KGM- I, KGM-II, KGM-III and konjac flour can lower blood sugar of alloxan-induced hyperglycemic mice.

Anti-Cancerous: Luo (1992) studied the inhibitory effect of refined *Amorphophallus konjac* (Konjaku powder) on MNNG-induced lung cancers and reported that refined *Amarphophallus konjac* powder caused a drop in cancer rate from 70.87% to 19.38% and reduction in mean number of cancer and precancerous lesions.It also altered the constituent ratio of the kinds of tumors, showing a decrease in malignancy (adenoma with malignant change), absence of adenocarcinoma, and relative increase in benign adenoma.

References

Angayarkanni, J., Ramkumar, K.M., Poornima, T., and Priyadarshini, U. 2007. Cytotoxic activity of *Amorphophallus paeniifolius* tuber extracts *in vitro*. American-Euresian J Agric Environ Sci., (2007):2:395–8.

Ansil, P.N., Prabha, S.P., Nitha, A., Wills, P.J., Jazaira, V., and Latha, M.S. 2012. Curative effect of *Amorphophallus campanulatus* (Roxb.) Blume. Tuber methanolic extract against thioacetamide induced oxidative stress in experimental rats. Asian Pacific Journal of Tropical Biomedicine (2012)S83-S89.

Das D., Mondal, S., Roy, S.K., Maiti, D., Bhunia, B., Maiti, T.K. and Islam, S.S. 2009. Isolation and characterization of a heteropolysaccharide from the corm of *Amorphophallus campanulatus*. Carbohydr Res., (2009), 344(18):2581-5.

Firdouse, S., and Alam, P. 2011. Phytochemical investigation of extract of *Amorphophallus campanulatus* tubers. International Journal of Phytomedicine, 3:32-35.

Khan, A., Rehman, M., and Islam, M.S. 2008. Antibacterial, antifungal and cytotoxic activities of amblyone isolated from *Amorphophallus campanulatus*. Indian J Pharmacol., (2008), 40(1):41-44.

Khare, C.P. 2007. Indian medicinal plants: an illustrated dictionary. New York: Springer; (2007), p. 45.

Li, C., Wang, Y., He, W., and Xie., B. 2004. Studies on the antidiabetic effect of konjac glucomannan with different molecular chains on experimental diabetes mice. Zhong Yao Cai., (2004), 27(2):110-3.

Luo, D. Y. 1992. Inhibitory effect of refined *Amorphophallus konjac* on MNNG-induced lung cancers in mice. Zhonghua Zhong Liu Za Zhi., (1992), 14(1):48-50.

Nair, R.V. 1993. Indian Medicinal Plants 1. Chennai: Orient Longman, (1993). Pp. 118–22.

Prathibha, S., Nambisan, B., and Leelamma, S. 1995. Enzyme inhibitors in tuber crops and their thermal stability., Plant foods. Hum Nutr., (1995), 48:247-57.

Sahu, K. G., Kjhadabadi, S. S., and Bhide, S. S. 2009. Evaluation of *in vitro* antioxidant activity of *Amorphophallus campanulatus* (Roxb.) Ex Blume Decne. Int. J. Chem. Sci., (2009), 7(3): 1553-1562.

Shilpi, J.A., Ray, P.K., Sardar, M.M., and Uddin, S.J. 2005. Analgesic activity of *Amorphophallus campanulatus* tuber. Fitoterapia, (2005),76:367–9.

Singh, S.K., Rajasekar, N., ArmstrongN., Raj, V., and Paramaguru, R. 2011. Hepatoprotective and antioxidant effects of *Amorphophallus Campanulatus* against acetaminopheninducedhepatotoxicity in rats. Int J Pharm Pharm Sci., 3(2): 202-205

6.6. Dioscorea Yam (*Dioscorea alata* L. or *Dioscorea esculanta* L)

Yam belongs to the family Dioscoreaceae and genus Dioscorea. In India *D. alata* and *D. esculanta* are extensively cultivated in large scale for edible purposes. *D.alata* is called the 'greater yam' which is round to oval or irregular in shape and the skin is black to brown in color, and flesh is white or yellowish or purplish. *D.esculenta* is called the 'lesser yam' which is thin and tubers are smaller as compared to other yams and are produced in clusters. They have thin yellowish-brown skin and white flesh. Another species *Dioscorea rotundata* (white yam or African yam) is widely under cultivation in Western Africa. It has been introduced into India and is gaining popularity.

Cultivars/Varieties: Three improved varieties of Greater Yam (*Dioscorea alata*) viz. Sree Keerthi, Sree Roopa and Sree Shilpa and three improved varieties of white yam (*Dioscorea rotundata*) viz. Sree Subhra, Sree Priya and Sree Dhanya were released form CTCRI, Thiruvananthapuram. Two varieties of lesser yam viz. Sree Latha and Sree Kala were also released from CTCRI.

Nutritional Components of *Dioscorea alata*

Constituents	Quantity (per 100 g)	Constituents	Quantity (per 100 g)
Moisture	70.4 g	Iron	1.0 mg
Carbohydrates	24.4 g	Phosphorus	74 mg
Proteins	2.5 g	Carotene	565 µg
Fats	0.3 g	Thiamine	0.19 mg
Energy	110 Kcal	Riboflavin	0.47 mg
Calcium	20 mg	Niacin	1.2 mg

Gopalan *et al.*, (1985)

Phytochemicals

The diascorea tubers contain nearly all the essential nutrients including minerals, vitamins and have several medicinal properties. They are rich sources of proteins and amino acids. Many species of yam contain small amounts of sapogenins and alkaloids. The main sapogenin present is diosgenin which is used in several corticosteroidal drugs. The main bioactive components of wild yam are the saponin, diosgenin, and the alkaloids, and dioscorine. These components are believed to have antispasmodic, cholagogue, and diaphoretic effects. Wild yam contains a progesterone precursor used by the pharmaceutical industry to produce progesterone. There are three types of additional phytohormones used for estrogen replacement therapy. Estriol, Estrone, and Estradiol are all derived from either soybean plants or Mexican wild yams.

Diosgenin was the main source of pharmaceutical corticosteroids and sex steroids during the 1950s. Those grown as steroidal sources or medicinal herbs are bitter, not commonly eaten, and their liquid extracts are internally or externally used to treat asthma, intestinal spasm, rheumatic pain, menopausal symptom and menstrual disorder in native Americans and in traditional Chinese medicine (Briggs, 1990; Moyad, 2002; Dentali, 1996; Taffe and Cauffield, 1998).

Health Benefits of Yam based on Traditional Knowledge and Folklore Practices

- Wild yam has been used as a form of traditional herbal Chinese medicine for the proper functioning of the renal and the female endocrine system.
- Wild yam also helps decrease water retention and alleviates nausea during pregnancy. Mexican wild yam is a very effective antispasmodic and is used in conditions like menstrual cramps, pain, improper circulation, neuralgia, nerve tension, muscle tension, abdominal and intestinal cramps and even in inflammatory stages of RA.
- Yams have high levels of Vitamin B6, which is a popular supplement for women suffering from depression due to premenstrual syndrome (PMS). Regular intake of Vitamin B6 is also beneficial against a number heart conditions and stroke, as they help in the breakdown of a chemical substance called homocysteine.
- The rate of conversion of the complex carbohydrates and fiber present in yam, into sugar is very slow and their absorption rate into the blood stream is also very slow. Thus, they help maintain the blood sugar level.
- Yams are rich in manganese, which is an essential factor for the metabolism of carbohydrate and antioxidant defenses. It is also a co-factor in a number of enzymes that are related to the production of energy.
- Because yams are a good source of potassium, it helps maintain the blood pressure level.
- The mucilaginous tuber milk contains allantoin, a cell-proliferant that speeds the healing process when applied externally to ulcers, boils, and abscesses. Its decoction is also used to stimulate appetite and to relieve bronchial irritation, cough *etc.*

Therapeutic Properties of Diascorea Yam

In Taiwan, yam is widely cultivated and used as tonic nourishment especially for postmenopausal women in recent years. The studies conducted by Wu *et al.*, (2005) showed that two third replacement of staple food with yam for 30 days increased serum estrone levels, which might benefit the declined estrogens in postmenopausal women. It is also proposed that the increase in serum SHBG levels, the decreases in serum free androgen index, urinary 16-hydroxyestrone, urinary isoprostane and plasma cholesterol levels, and the prolonged lag time of LDL oxidation, might potentially protect postmenopausal women against the risk of breast cancer and cardiovascular diseases.

Anti-diabetic: Maithili *et al.*, (2011) studied the effect of ethanolic extract of *Dioscorea alata* on diabetes in rats and noticed significant antidiabetic activity. Serum lipid levels, total protein, albumin, and creatinine were reversed toward near normal in treated rats as compared to diabetic control.

Anti-Osteoporotic: In studies on anti-osteoporotic activity of the ethanol extract of the rhizomes of *Dioscorea alata* L., the results suggested that it has a novel mechanism that drives the lineage-specific differentiation of bone marrow stromal cells. They found increase in the activity of alkaline phosphatase (ALP) and bone nodule formation in primary bone marrow cultures (Peng *et al.*, 2011).

Menopausal Symptoms: In treatment of menopausal symptoms, compared with placebo, *Diascorea alata* improved symptoms, particularly the psychological parameters in menopausal women. Improvements were noted in the parameters like feeling tense or nervous, insomnia, excitable and musculoskeletal pain among those receiving Diascorea. Safety monitoring indicated that standardized extracts of *Diascorea alata* were safe during daily administration over a period of 12 months. (Hsu *et al.*, 2011). Diascorea consumption also resulted in positive effects on blood hormone profiles.

Immunomodulation: Yam dioscorin is also reported to function as an immunomodulatory substance (Liu *et al.*, 2007). *Dioscorea alata* L. feeding exhibited its antioxidative effects in hyperhomo-cysteinemia (Hhcy) including alleviating lipid peroxidation, and oxidative stress (Chang *et al.*, 2004). Dioscorin and its hydrolysates might be a potential for hypertension control when people consume yam tuber (Hsu *et al.*, 2002).

Good for Kideny & Liver: The pharmacological and biochemical studies showed the extract of yam had the effect of kidney secureness and liver fortification. The pathologic sections showed good improvements in renal tubular degranulation changes, necrosis and disintegration. The extract of yam also possessed a good protection against the inflammation of central vein and necrosis of liver tissue, which suggests that the yam could prevent the damages of the liver and kidneys, thus preserving their functions (Lee *et al.*, 2002).

Note: Yams of African species must be cooked before safely eaten, because various natural toxin substances such as dioscorine can cause illness if consumed raw.

References

Briggs, C.J. 1990. Herbal medicine: Dioscorea: The yams—A traditional source of food and drugs. Can Pharm J., 123:413–415.

Chang, S.J., Lee, Y.C., Liu, S.Y., and Chang, T.W. 2004. Chinese yam (*Dioscorea alata* cv. Tainung No. 2) feeding exhibited antioxidative effects in hyperhomocysteinemia rats. J Agric Food Chem., (2004), 52(6):1720-5.

Dentali, S. 1996. Clearing up confusion over yams and progesterone. Altern Ther Health, M 2:19–20.

Hsu, C.C., Kuo, H.C., Chang, S.Y., Wu, T.C., and Huang, K.E. 2011.The assessment of efficacy of *Diascorea alata* for menopausal symptom treatment in Taiwanese women. Climacteric., (2011), 14(1):132-9.

Hsu, F.L., Lin, Y.H., Lee, M.H., Lin, C.L., and Hou, W.C. 2002. Both dioscorin, the tuber storage protein of yam (*Dioscorea alata* cv. Tainong No. 1), and its peptic hydrolysates exhibited angiotensin converting enzyme inhibitory activities. J Agric Food Chem., (2002), 50(21):6109-13.

http://lifestyle.iloveindia.com/lounge/benefits-of-yams-5758.html

http://www.althealth.co.uk/help-and-advice/supplements/yam/

Lee, S.C., Tsai, C.C., Chen, J.C., Lin, J.G., Lin, C.C., Hu, M.L., Lu, S. 2002. Effects of Chinese yam on hepato-nephrotoxicity of acetaminophen in rats. Acta Pharmacol Sin., (2002), 23(6):503-8.

Liu, Y.W., Shang, H.F., Wang, C.K., Hsu, F.L., and Hou, W.C. 2007. Immunomodulatory activity of dioscorin, the storage protein of yam (*Dioscorea alata* cv. Tainong No. 1) tuber. Food Chem Toxicol., (2007), 45(11):2312-8.

Maithili, V., Dhanabal, S.P., Mahendran, S., and Vadivelan, R. 2011. Antidiabetic activity of ethanolic extract of tubers of *Dioscorea alata* in alloxan induced diabetic rats. Indian J Pharmacol., (2011), 43(4):455-9.

Moyad, M.A. 2002. Complementary/alternative therapies for reducing hot flashes in prostate cancer patients: reevaluating the existing indirect data from studies of breast cancer and postmenopausal women. Urology, (2002), 59:20–33.

Peng, K.Y., Horng, L.Y., Sung, H.C., Huang, H.C., and Wu, R.T. 2011. Antiosteoporotic activity of *Dioscorea alata* L. cv. Phyto through driving mesenchymal stem cells differentiation for bone formation. Evid Based Complement Alternat Med., (2011), Volume 2011, Article ID 712892. Doi:10.1155/2011/712892

Taffe, A.M., and Cauffield, J. 1998. Natural hormone replacement therapy and dietary supplements used in the treatment of menopausalsymptoms. Lippincott's Prim Care Pract., (1998), 2:292–302.

Wu, W.H., Liu, L.Y., Chung, C.J., Jou, H.J., and Wang, T.A. 2005. Estrogenic effect of yam ingestion in healthy postmenopausal women. J Am Coll Nutr., (2005), 24(4):235-43.

7. Mallow Vegetables

7.1. Okra or Bhindi (*Abelmoschus esculentus* L.)

Okra (Bhindi) is a member of Malvaceae family, earlier known as *Hibiscus esculentus,* and is an annual vegetable crop grown from seeds in tropical and subtropical parts of the world. Okra, also known as *"lady finger"* or "gumbo", is a highly nutritious green edible pod vegetable. Botanically, this perennial flowering plant belongs to the *mallow* family but is grown as an annual. It is named scientifically as *Abelmoschus esculentus.* Okra originated in tropical and subtropical Africa and India is suggested to be the secondary centre of origin. There are four species under Abelmoschus genus which includes both cultivated and wild forms. They are *A. esculentus, A. manihot, A. caillei and A. moschatus.* In India, Okra is widely cultivated in states of Andhra Pradesh, West Bengal, Orissa, Bihar Uttar Pradesh, Karnataka, Assam West Bengal, Orissa, Gujarat and Bihar.

Cultivars/Varieties: Okra varieties differ in growth habit, height of plant, presence of purple pigmentation and ridges on fruit. The popular varieties under cultivation in different parts of the country are Arka Abhay, Arka Anamika, Azad Kranti, Co 1, MDU1, Gujarat Bhindi 1, Harbhajan Bhindi, Hisar Unnat, Parbhani Kranti, Perkins Long Green, Punjab 7, Punjab 8, Punjab Padmini, Pusa A 4, Pusa Makhmali, Pusa Sawani, Red Bhindi, TN Hybrid 8 and Varsha Uphar. Besides, a number of hybride from many private sector agencies/seed companies are marketed. Of these, Varsha, Vijay, Adhunik, Panchali, Hybrid No. 6,7 and 8, Nath Sobha, Supriya, Sungro 35, Aroh 1, - 2, -3, -9 and are more popular.

Uses

Different parts of the plant are used for different purposes. *Abelmoschus moschatus* Medik. Leaves and seeds are considered as valuable traditional medicine. *Abelmoschus moschatus* is also called 'musk okra' a medicinal plant belonging the same family. The leaves and new shoots of this plant is also used as vegetable. The aromatic seeds of this plant are aphrodisiac, ophthalmic, cardio tonic, antispasmodic and used in the treatment of and intestinal complaints and anxiety.

Nutritional Components of Okra

Constituents	Quantity (per 100 g)	Constituents	Quantity (per 100 g)
Moisture		Calcium	81 mg
Carbohydrates	7.03 g	Potassium	303 mg
Proteins	2.0 g	Vitamin A	371 IU
Fats	0.1 g	Thiamine	0.200 mg
Energy	31 Kcal	Riboflavin	0.060 mg
Dietary Fibre	3.2 g	Niacin	1.00 mg

Constituents	Quantity (per 100 g)	Constituents	Quantity (per 100 g)
Iron	0.80 mg	Vitamin E	0.36 mg
Phosphorus	63 mg	Vitamin K	53 mcg

Gopalan *et al.*, (1985)

Phytochemicals

Nutritionally, there are many useful substances in the seeds of *A. esculentus*, such as, flavones, polysaccharides, pectin, trace elements, and amino acids. Water extracts of *A. esculentus* showed presence of phenolics and flavonoids (Adelakun *et al.*, 2009; Huang *et al.*, 2007). *A. esculentus* could be the potentially rich source of natural antioxidant as indicated by its evaluation using leaf, fruit, flower and seed extracts (Liao *et al.*, 2012). A new flavonol glycoside characterized as 5,7,3',4'-tetrahydroxy-4''-O-methyl flavonol -3-O-β-D- glucopyranoside (1) was isolated from the fruit of *A. esculentus* together with one known compound 5,7,3',4'-tetrahydroxy flavonol -3-O-[β-D-glucopyranosyl-(1→6)]-β-D-glucopyranoside (2) (Liao *et al.*, 2012).

Okra is treasure of polyphenolic substances. The major polyphenols documented in okra are hyperoside, quercetin, coumarin scopoletin, uridine, and phenylalanine (Bandyukova & Ligai, 1987; Lu, Huanfen, & Linlin, 2011). Shui & Peng (2004) have reported that quercetin derivatives and (−)-epigallocatechin as major antioxidant compounds in okra. 70% of the total antioxidant activity comes due to the quercetin derivatives (quercetin 3-O-xylosyl (1''→2'') glucoside, quercetin 3-O-glucosyl (1''→6'') glucoside, quercetin 3-O-glucoside and quercetin 3-O- (6''-O-malonyl)-glucoside).

Procycanidin B2 was found to be the predominant phenolic compound followed by procycanidin B1 and rutin in seeds. In pulped seed catechin, procycanidin B2, epicatechin and rutin are reported to be present (Khomsug *et al.*, 2010).

Okra pods are mucilaginous, low in calories but nutritionally rich and a good source of edible fiber. Studies have shown that the okra pod contains important bioactive compounds such as carotene, folic acid, thiamine, riboflavin, niacin, vitamin C, oxalic acid and amino acids. (Roy *et al.*, 2014). In Ayurveda, okra is used as an edible infusion and in different preparation for diuretic effect (Maramag, 2013). An infusion of the fruit mucilage is also used to treat dysentery and diarrhoea in acute inflammation and irritation of the stomach, bowels, and kidneys catarrhal infections, ardour urinae, dysuria and gonorrhoea. Seeds are used as antispasmodic, cordial and stimulant (Lim, 2012). Leaves and root extracts are served as demulcent and emollient poultice (Babu & Srinivasan, 1995).

Health Benefits of Okra Based on Traditional Knowledge and Folklore Practices

- The soluble fibre in okra helps to lower serum cholesterol, reducing heart diseases. The water insoluble fibres helps to keep the intestinal tract healthy, decreasing the risk of some forms of cancer, especially colorectal cancer.
- The fruits as well as roots of the plants contain mucilages and have strong demulcent action. This mucilage can be used as plasma replacement.

- An infusion of the roots is used in the treatment of syphilis. The leaves are used as an emollient poultice.
- Decoction of immature capsules (pods) is demulcent, diuretic and emollient, and is used in the treatment of catarrah infections, ardor urinae, ioavai and gonorrhea.
- The seeds are antispasmodic, cordial, and stimulant.
- An infusion of the roasted seeds has sudorfic properties.
- Tuberous roots of wild okra are used to overcome male infertility by tribals in India.
- It is very low in calories, but is a rich source of dietary fiber, minerals, vitamins; recommended in cholesterol controlling and weight reduction programs.
- The pods contain healthy amounts of vitamin A, and flavonoid antioxidants such as beta carotenes, xanthin and lutein.
- Fresh pods are good source of folates; provide about 22% of RDA per 100 g.
- The pods are also an excellent source of antioxidant vitamin, vitamin-C;
- They are rich in B-complex group of vitamins like niacin, vitamin B-6 (pyridoxine), thiamin and pantothenic acid. The pods also contain good amounts of vitamin K. Vitamin K is a co-factor for blood clotting enzymes and is required for strengthening of bones.
- The pods are also good source of many important minerals such as iron, calcium, manganese and magnesium.

Therapeutic Properties of Okra

Anti-diabetic & Anti-hyperlipidemia: The study of antidiabetic and anti-hyperlipidemic potential of *Abelmoschus esculentus* peel and seed powder (AEPP and AESP) in streptozotocin (STZ)-induced diabetic rats showed that elevated lipid profile levels returned to near normal in diabetic rats after the administration of AEPP and AESP at 100 and 200 mg/kg dose, compared to diabetic control rats (Savitha *et al.*, 2011).

The 95% ethanol extract from the flowers of *Abelmoschus ioavai* (L.) Medic showed inhibitory activity on Triglycerides (TG) accumulation in 3T3-L1 preadipocyte. Chemical studies on the active fraction led to the isolation of 14 flavonoids. Ten of these 14 flavonoids showed inhibitory activity on TG accumulation significantly in mature 3T3-L1 cells (An *et al.*, 2011). This shows its effectiveness in regulating lipid metabolism by liver.

Anti-microbial & Antioxidant: To give a scientific basis for traditional usage of this medicinal plant, the seed and leaf extracts were evaluated for their antioxidant, free radical scavenging, antimicrobial and antiproliferative activities (Gul *et al.*, 2011). The seed and leaf extracts of *A. moschatus* possess significant antioxidant activity and could serve as free radical inhibitors or scavenger, probably as primary antioxidants. The investigations showed that plant possesses moderate antibacterial activity against bacterial strains (*Bacillus subtilis and Staphylococcus aureus, Escherichia coli, Pseudomonas aeruginosa, Proteus vulgaris and Salmonella enterica paratyphi*) and one fungal strain (*Candida albicans*).

Anti-cancerous: Hydroalcoholic seed and leaf extracts also exhibited anti-proliferative activity against two human cancer cell lines (colorectal adenocarcinoma and retinoblastoma human cancer cell lines). *A. moschatus* may therefore, be a good candidate for functional foods as well as pharmaceutics. The Okra seeds are also reported to have antiproliferative activity. The proliferation and apoptosis of metastatic melanoma cells are often abnormal. Vayssade *et al.*, (2010) evaluated the action of a pectic rhamnogalacturonan obtained by hot buffer extraction of okra pods (okra RG-I) on melanoma cell growth and survival *in vitro*. Their findings suggest that okra RG-I induces apoptosis in melanoma cells by interacting with Gal-3. As these interactions might open the way to new melanoma therapies, the next step will be to determine just how they occur.

Neuroprotective: The *Abelmoschus esculentus* is believed to have neuroprotective effects. A study aimed to examine the ability of okra (*Abelmoschus esculentus* Linn.) extract and its derivatives (quercetin and rutin) to protect neuronal function and improve learning and memory deficits in mice subjected to dexamethasone treatment was examined in Thailand (Tongjaroenbuangam *et al.*, 2011) which suggest that quercetin, rutin and okra extract treatments reversed cognitive deficits, including impaired dentate gyrus (DG) cell proliferation, and protected against morphological changes in the CA3 region in dexamethasone-treated mice.

Anti-Osteoporotic: The biggest culprit in pathogenesis of osteoporosis, particularly in postmenopausal women is oestrogen deficiency. Hormone replacement therapy remained the mainstay for prevention of osteoporosis in post menopausal women. Puel *et al.*, (2005) investigated the ability of *Abelmoschus moschatus*, to prevent bone loss in ioavailabled rats. *Abelmoschus moschatus*, produces large leaves and flowers. While the leaves are used as vegetable, its root is also used in folklore medicine. The findings of Puel *et al.*, (2005) suggested that *Abelmoschus moschatus* consumption, at the dose of 15% in the diet, provided bone-sparing effects by improving both BMD (Bone Mineral Density) and BMC (Bone Mineral Content).

Kidney Diseases: Nephropathy is one of the renal complications associated with non-insulin dependent diabetes mellitus (NIDDM). The *Abelmoschus moschatus* is believed to help in improving the renal function, so human volunteers with this complication were treated with alcohol extraction of *Abelmoschus moschatus*, Gliclazide and Captopril tablets and control. The results of the study suggested that *Abemoschus moschatus* alcohol extraction could eliminate oxygen free radicals, alleviate renal tubular-interstitial diseases, improve renal function and reduce proteinuria (Yu *et al.*, 1995).

References

Adelakun, O.E., Oyelade, O.J., Ade-Omoway, B.I., Adeyemi, I.A., Van de Venter, M. 2009. Chemical composition and the antioxidative properties of Nigerian okra seed (*Abelmoschus esculentus* Moench) flour. J Food Chem Toxicol., (2009), 47:1123–6.

An, Y., Zhang, Li C,Y., Qian, Q., He, W., and Wang, T. 2011. Inhibitory effects of flavonoids from *Abelmoschus ioavai* flowers on triglyceride accumulation in 3T3-L1 adipocytes. Fitoterapia, (2011), 82(4):595-600.

Babu, P. S., and Srinivasan, K. 1995. Influence of dietary curcumin and cholesterol on the progression of experimentally induced diabetes in albino rat. Molecular and Cellular Biochemistry, 152: 13– 21.

Bandyukova, V. A., and Ligai, L. V. 1987. A chemical investigation of the fruit of *Abelmoschus esculentus*., Chemistry of natural compounds, 23: 376-377. http://dx.doi.org/10.100/BF00600851.

Dhankar, B.S. and Singh, R. 2010. Okra Handbook: Global Production, Processing, and Crop Improvement. Pp.94.

Gul, M.Z., Bhakshu, L.M., Ahmad, F., Kondapi, A.K., Qureshi, I.A., Ghazi, I.A. 2011. Evaluation of *Abelmoschus moschatus* extracts for antioxidant, free radical scavenging, antimicrobial and antiproliferative activities using *in vitro* assays. BMC Complement Altern Med., (2011), 17:11:64.

Huang, A.G., Chen, X.H., Gao, Y.Z., and Che, J. 2007. Determination and analysis of ingredient in okra. Chin J Food Sci., (2007), 28:451–5.

Khomsug, P., Thongjaroenbuangam, W., Pakdeenarong, N., Suttajit, M., and Chantiratikul P. 2010. Antioxidative activities and phenolic content of extracts from Okra (*Abelmoschus esculentus* L.) Research Journal of Biological Sciences, (2010), 5(4): 310-313.

Liao, H., Dong, W., Shi, X., Liu, H., and Yuan, K. 2012. Analysis and comparison of the active components and antioxidant activities of extracts from *Abelmoschus esculentus* L. Pharmacogn Mag., (2012), 8(30):156-61.

Liao, H., Liu, H., and Yuan, K. 2012. A new flavonol glycoside from the *Abelmoschus esculentus* Linn. Pharmacogn Mag., (2012), 8(29):12-5.

Lim T. K. 2012. Edible Medicinal and Non-Medicinal Plants, 3:160. http://dx.doi.org/10.1007/978-94-007-2534-8_21.

Lu, J., Huanfen, L., and Linlin, J. 2011. Chemical constituents in n-butanol extract of Abelmoschus esculentus. Chinese Traditional and Herbal Drugs, 41: 1771-1773.

Maramag, R. P. 2013. Diuretic potential of *Capsicum frutescens* L., *Corchorus oliturius* L., and *Abelmoschus esculentus* L. Asian Journal of Natural And Applied Science, (2013), 2 (1): 60-69.

Puel, C., Mathey, J., Kati-Coulibaly, S., Davicco, M.J., Lebecque, P., Chanteranne, B., Horcajada, M.N., and Coxam, V. 2005. Preventive effect of *Abelmoschus ioavai* (L.) Medik. On bone loss in the ioavailabled rats. J Ethnopharmacol., (2005), 99(1):55-60.

Roy, A., Shrivastava, S.L., Mandal, S.M., Santi, M. 2014. Functional properties of okra *Abelmoschus Esculentus* L. (Moench): Traditional Claims and Scientific Evidences. Plant Science Today, 1(3): 121-130. Doi.org/10.14719/Pst.2014.1.3.63

Sabitha, V., Ramachandran, S., Naveen, K.R., and Panneerselvam, K. 2011. Antidiabetic and antihyperlipidemic potential of *Abelmoschus esculentus* (L.) Moench. In streptozotocin-induced diabetic rats. J Pharm Bioallied Sci., (2011), 3(3): 397–402.

Shui, G., and Peng, L. L. 2004. An improved method for the analysis of major antioxidants of Hibiscus esculentus Linn. Journal of Chromatography A, 1048: 17–24.

Tongjaroenbuangam, W., Ruksee, N., Chantiratikul, P., Pakdeenarong, N., Kongbuntad, W., and Govitrapong, P. 2011. Neuroprotective effects of quercetin, rutin and okra (*Abelmoschus esculentus* Linn.) in dexamethasone-treated mice. Neurochem Int., (2011), 59(5):677-85.

Vayssade, M., Sengkhamparn, N., Verhoef, R., Delaigue, C., Goundiam, O., Vigneron, P., Voragen, A.G., Schols, H.A., and Nagel, M.D. 2010. Antiproliferative and proapoptotic actions of okra pectin on B16F10 melanoma cells. Phytother Res., (2010), 24(7):982-9.

Yu, J.Y., Xiong, N.N., and Guo, H.F. 1995. Clinical observation on diabetic nephropathy treated with alcohol of *Abelmoschus ioavai*. Zhongguo Zhong Xi Yi Jie He Za Zhi., (1995), 15(5):263-5.

8. Leguminous Vegetables

Leguminous vegetables include peas, French bean, Lima bean, broad bean, cowpea, hyacinth bean, winged bean *etc*. They belong to the family Fabaceae (Syn. Leguminosae). These are cultivated either as sole crop is sometimes as intercrop. In some areas they are used in backyard cultivation also. Most of them are short duration crops and are self pollinated.

8.1. Peas (*Pisum sativum* L.)

The pea (Pisum *sativum* L.) is a very common nutritioius vegetable grown in cool season throughout the world. It belongs to the family Fabaceae (Syn. Leguminaceae) and genus Pisum. A pea is most commonly the small spherical seed or the seedpod of the pod fruit *Pisum sativum* L. The pea cultivars are grouped into several groups based on shape of seed, plant height, maturity and use of pods. This crop is cultivated for its tender and immature pods for use as vegetable and mature dry pods for use as a pulse. In both the cases the seeds are separated and used as vegetable or pulse. The central Asia is regarded as the place of origin for all legumes including pea. Based on the genetic diversity, Central Asia, the Near East, Abyssinia and the Mediterranean is considered to be its centre of Origin. Cultivated garden pea is not seen in wild state and it might have been originated from wild field pea or other related species. In India it widely cultivated states of Uttar Pradesh, Bihar, Haryana, Punjab, Himachal Pradesh, Orissa and Karnataka. Followed by Madhya Pradesh, Himachal Pradesh, Assam and Jharkhand.

Cultivars/Varietie: Peas varieties are grouped based on maturity (early, mid season and late) plant height (Bush or dwarf, medium tall and tall). The major cultivars under cultivation in different parts of the country are Arkel, Bonneville, Harbhajan, FC-1, Arka Ajit, UN 53-6 (whole edible type), Jawahar Matar 1, Jawahar Matar 2, Jawahar Matar 3, Jawahar Matar 4, Jawahar Matar 5, Jawahar Peas 83, JP 4 (JM 6), JP 19, Lincoin, Mattar Ageta 6, Pantnagar Matar 2, Pant Uphar (IP 3), P 88, PRS 4,Ooty-1 and VL 3.

Uses

Fresh peas are used primarily used as vegetable, in preparation various kinds of dishes. Often it is used in combination with other vegetables like potato, cauliflower, carrot and also with cheese/paneer. They are also used as stuffing or filling in stuffed paratha/bread. The green peas are frozen or canned for long term use or exports. The dreid peas are rehydrated and used as substitute for green peas during off season. It is also extensively used in Indian savory industry for making various kinds of snack foods.

Being a leguminous vegetable, it is rich in proteins and fibre.

Nutritional Components of Peas

Constituents	Quantity (per 100 g)	Constituents	Quantity (per 100 g)
Moisture	72.0 g	Calcium	20 mg
Carbohydrates	15.8 g	Potassium	79 mg
Proteins	7.2 g	Magnesium	34 mg
Fats	0.1 g	Vitamin A	139 IU
Energy	93 Kcal	Thiamine	0.25 mg
Dietary Fibre	4.0 g	Riboflavin	0.01 mg
Iron	1.5 mg	Niacin	0.80 mg
Phosphorus	139 mg	Vitamin C	9 mg

Gopalan, *et al.*, (1985)

Peas are rich in proteins, fibres, iron, phosphorus, magnesium and potassium. The mineral availability from legume vegetables differs and may be attributed to their mineral content, mineral-mineral interaction and from their phytic and tannic acid content. Lgumes are considered low-GI foods and have shown potential hypocholesterolaemic effects.

Phytochemcials

In the flavonoid category, green peas contain antioxidants catechin and epicatechin. In the carotenoid category, they offer alpha-carotene and beta-carotene. Their phenolic acids include ferulic and caffeic acid. Their polyphenols include coumestrol. Pisum saponins I and II and pisomosides A and B are anti-inflammatory phytonutrients found almost exclusively in peas. Recent research has shown that green peas are a reliable source of omega-3 fatty acid in the form of alpha-linolenic acid (ALA).

Two new anthocyanins were isolated from purple pods of pea (*Pisum spp.*). Their structures were identified as delphinidin 3-xylosylgalactoside-5-acetylglucoside and its deacetylated derivative by the usual chemical degradation methods and by spectroscopic methods such as UV-VIS, MS and NMR. Both pigments showed moderate stability and antioxidative activity in a neutral aqueous solution.

The antioxidant and anti-inflammatory properties of peptides from yellow field pea proteins (*Pisum sativum* L.) were investigated which showed that enzymatic protein degradation confers antioxidant, anti-inflammatory and immunomodulating potentials to pea proteins, and the resultant peptides could be used as an alternative therapy for the prevention of inflammatory-related diseases (Ndiaye *et al.*, 2011).

In comparison to glutathione, the peptide fractions from pea protein hydrolysates had significantly higher ($p < 0.05$) ability to inhibit linoleic acid oxidation and chelate metals. In contrast, glutathione had significantly higher ($p < 0.05$) free radical scavenging properties than the peptide fractions (Pownall *et al.*, 2010).

Epidemiologic studies have shown that women with a higher dietary intake of phytoestrogens, plant-derived compounds with partial estrogen agonist properties, have a lower incidence of cardiovascular disease and breast and uterine cancer than women with a lower dietary intake of these substances.

Health Benefits of Consuming Garden Peas

- Peas are one of the most nutritious leguminous vegetable, rich in health benefiting phyto-nutrients, minerals, vitamins and antioxidants.
- Peas are relatively low in calories when compared with beans, and cowpeas. It contains good amount of soluble and insoluble fiber and contains no cholesterol.
- Fresh pea pods are excellent source of folic acid. Folates are B-complex vitamins required for DNA synthesis inside the cell. It has been well established through research studies that adequate folate rich foods in expectant mothers would help prevent neural tube defects in the newborn babies.
- Fresh green peas have good amounts of ascorbic acid (vitamin C). Vitamin C is a powerful natural water-soluble antioxidant. Vegetables rich in this vitamin helps body develop resistance against infectious diseases and scavenge harmful, pro-inflammatory free radicals from the body.
- Peas contain phytosterols especially ß-sitosterol. Studies suggest that vegetables like legumes, fruits and cereals rich in plant sterols help lower cholesterol levels in the body.
- Garden peas are also good in vitamin K. 100 g of fresh leaves contain about 24.8 mcg or about 21% of daily requirement of vitamin K-1 (phylloquinone). Vitamin K has found to have potential role in bone mass building function by promoting osteo-trophic activity in the bone. It also has established role in Alzheimer's disease patients by limiting neuronal damage in the brain.
- Fresh green peas also contain adequate amounts of antioxidants flavonoids such as carotenes, lutein and zeaxanthin as well as vitamin-A (provide 765 IU or 25.5% of RDA per 100 g). Vitamin A is essential nutrient which is required for maintaining healthy mucus membranes and skin and is also essential for vision. Consumption of natural fruits rich in flavonoids helps to protect from lung and oral cavity cancers.
- In addition to folates, peas are also good in many other essential B-complex vitamins such as pantothenic acid, niacin, thiamin, and pyridoxine. They are also rich source of many minerals such as calcium, iron, copper, zinc and manganese.
- The critical cardioprotective B vitamin, choline, is also provided by green peas. In combination, these nutrient features of green peas point to a likely standout role for this food in protection of our cardiovascular health.
- A study conducted in Mexico City has shown that daily consumption of green peas along with other legumes is associated with decreased risk of stomach cancer. Average daily intake of a polyphenol called coumestrol at a level of 2 milligrams or higher decreased risk of stomach cancer.

References

Ndiaye, F., Vuong, T., Duarte, J., Aluko, R.E., and Matar, C. 2011. Antioxidant, anti-inflammatory and immunomodulating properties of an enzymatic protein hydrolysate from yellow field pea seeds. Eur J Nutr., (2012), 51(1):29-37.

Pownall, T.L., Udenigwe, C.C., and Aluko, R.E. 2010. Amino acid composition and antioxidant properties of pea seed (*Pisum sativum* L.) enzymatic protein hydrolysate fractions. J Agric Food Chem., (2010), 58(8):4712-8.

Terahara, N., Honda, T., Hayashi, M., and Ishimaru. K. 2000. New anthocyanins from purple pods of pea (*Pisum spp.*). Biosci Biotechnol Biochem., (2000), 64(12):2569-74.

Xu, B.J., Yuan, S.H., and Chang, S.K. 2007. Comparative studies on the antioxidant activities of nine common food legumes against copper-induced human low-density lipoprotein oxidation *in vitro*. J Food Sci., (2007), 72(7): S522-7.

8.2. French Bean (*Phaseolus vulgaris* L.)

French bean (*Phaseolus vulgaris* L.) is also known as common bean, kidney bean, dwarf bean, snap bean, string bean or garden bean. French bean is an important legume vegetable grown for its tender pods, shelled green beans and dry beans.

French bean has originated in Central and South America. In India they are widely grown in states of West Bengal, Andhra Pradesh, Maharashtra, Jharkhand, Karnataka and Orissa.

Cultivars/Varieties: In India beans are mostly grown for tender vegetable, while in USA it is grown for processing in large quantities. There are specific varieties for snap bean purpose, dry bean purpose and for processing. There are a large number of varieties under cultivation. Important ones are Arka Komal, Arka Suvidha, Arka Sukomal, Pusa Parvati, Pusa Himlatha, Ooty-1, KKL-1, Phule Surekha, Pant Anupama, NDVP 8 & 10.

Uses

French bean pods are primarily used for vegetable purpose wholly. While domestically the fresh beans are preferred, large quantities of French beans are processed (dehydrated) and exported from India. It is a nutritive vegetable and is grown under varied cropping patterns in hills and in southern region of India. It is nutritious vegetable rich in proteins, calcium, phosphorus and iron.

Nutritional ioavailab of French bean

Constituents	Quantity (per 100 g)	Constituents	Quantity (per 100 g)
Moisture	91.4	Potassium	129 mg
Carbohydrates	4.50 g	Sulphur	37 mg
Proteins	1.70 g	Vitamin A	221 IU
Fats	0.10 g	Thiamine	0.08 mg
Energy	31 Kcal	Riboflavin	0.06 mg
Dietary Fibre	1.80 g	Nicotinic acid	0.30 mg
Iron	1.70 mg	Vitamin C	16.00 mg
Phosphorus	28.00 mg	Folic acid	734 mcg
Calcium	50 mg	Vitamin B6	738 mcg

Gopalan, *et al.*, (1985)

The beans are said to be good for bladder burns, cardiac, carmative, depurative, diarrhea, diuretic, dropsy, dysentery, eczema, emollient, hiccups, itch, kidney resolvent, rheumatism, sciatica and tenesmus (Duek, 1981).

Phytochemicals

Antioxidants are important in protection against hypertension, diabetes, cardiovascular diseases and cancer. Green beans contain a wide variety of carotenoids (including lutein, beta-carotene, violaxanthin, and neoxanthin) and flavonoids (including quercetin, kaemferol, catechins, epicatechins, and procyanidins) that have all been shown to have health-supportive antioxidant properties. Among the leguminous vegetables, string beans have the highest antioxidant capacity compared to the others because of significantly higher levels of total phenolic, ascorbic acid and ß-carotene contents (Amin *et al.*, 2009).

Health Benefits of consuming French beans

- The antioxidant present in beans help the body build resistance against infectious diseases and scavenge the reactive oxygen species that play a role in the aging process and lifestyle diseases.

- As French beans possess good quantity dietary fiber, it acts as a bulking agent or laxative that helps to protect the mucus membrane of the colon by decreasing its exposure to toxic substances. Dietary fiber is also known to reduce blood cholesterol levels by decreasing re-absorption of cholesterol binding acids in the colon.

- Zeaxanthin in French beans is absorbed into the retinal macula in the eyes and provides antioxidant and light filtering functions. It is known to be helpful in preventing age related macular diseases.

- The folate helps in preventing neural tube defects in the growing foetus and prevents the accumulation of an intermediary metabolite of protein metabolism, called homocysteine, which promotes the risk of atherosclerosis.

- French beans are a good source of molybdenum that helps in detoxification of sulfites from the blood.

- Copper, found in French beans, is known to lower the risk of inflammatory diseases, like rheumatoid arthritis and maintain the elasticity of blood vessels, joints and ligaments by enhancing the activity of the enzymes.

- The magnesium helps in relieving fatigue, relaxing sore muscles, nerves and blood vessels, thereby relieving the symptoms of asthma and migraine headaches.

- The soluble fiber slows down the metabolism of carbohydrates which, in turn, regulates the blood sugar levels and prevents a sudden jump in blood sugar levels after meals. It is good for diabetic people and those with insulin resistance.

Therapeutic Properties of French Bean

Anti-diabetic: French beans in diets has been found benefit in weight reduction and blood sugar management. As an alternative to a low glycemic index diet, there is a growing body of research into products that slow the absorption of carbohydrates through the inhibition of enzymes responsible for their digestion. These products include alpha-amylase and glucosidase inhibitors. The common white bean (*Phaseolus vulgaris*) produces an alpha-amylase inhibitor, which has been characterized and tested in numerous clinical studies (Berrett and Udani, 2011). Although a number of natural supplements with anti-amylase activity have been recognized, the most studied and favored one is white kidney bean extract. Animal and human studies clearly show that this agent works *in-vivo* and has clinical utility (Obiro *et al.*, 2008; Preuss, 2009, Helmstadter, 2010). In a study conducted by Celleno *et al.*, (2007), after 30 days, subjects receiving *Phaseolus vulgaris* extract with a carbohydrate-rich, 2000- to 2200-calorie diet had significantly (p<0.001) greater reduction of body weight, BMI, fat mass, adipose tissue thickness, and waist,/hip/ thigh circumferences while maintaining lean body mass compared to subjects receiving placebo.

Beans have also been known to have significant effect of reducing the blood sugar and serum cholesterol levels in diabetics. Pari and Venkateswaran (2004) conducted an investigation to evaluate the effect of *Phaseolus vulgaris*, on blood glucose, plasma insulin, cholesterol, triglycerides, free fatty acids, phospholipids, and fatty acid composition of total lipids in liver, kidney, and brain of normal and streptozotocin (STZ) diabetic rats. Their results suggest that *Phaseolus vulgaris* pod extract (PPEt) exhibits hypoglycemic and hypolipidemic effects in STZ diabetic rats. It also prevents the fatty acid changes produced during diabetes. The effect of PPEt at 200 mg/kg of body weight was better than that of glibenclamide drug. It significantly reduced the elevated blood glucose, serum triglycerides, free fatty acids, phospholipids, total cholesterol, very-low-density lipoprotein cholesterol, and low-density lipoprotein cholesterol (Venkateswaran *et al.*, 2002).

Anti-Hyperlipidemia: It is also widely reported that baked or cooked beans have cholesterol lowering effect. Costa *et al.*, (1993) studied the effect of graded inclusion of baked beans (*Phaseolus vulgaris*) on plasma and liver lipids in hypercholesterolaemic pigs fed on a Western-type diet and found that supplements of 200 and 300 g baked beans/kg promoted a significant (P < 0.05) reduction of about 50% in cholesterol deposition in the liver, compared with the control. To investigate the effective portion of black beans (*Phaseolus vulgaris)* on the cholesterol lowering effect, Rosa *et al.*, (1998) fed hypercholesterolemic rats with black beans without hulls and concluded that the hypocholesterolemic compounds of beans seem to be located in the inner part of the grain.

Anti-Cancerous: Emerging evidence indicates that common bean (*Phaseolus vulgaris* L.) is associated with reduced cancer risk in human populations and rodent carcinogenesis models. The role of lectins in prevention of proliferation of malignant cells have been well established. Chan *et al.*, (2012) found that a dimeric 64-kDa glucosamine-specific lectin purified from seeds of *Phaseolus vulgaris* cv. "brown kidney bean" exhibited antiproliferative activity toward human breast cancer (MCF7) cells, hepatoma (HepG2) cells and nasopharyngeal carcinoma (CNE1 and CNE2) cells. Cell signaling pathways associated with a reduction in mammary cancer burden by dietary common bean (*Phaseolus vulgaris* L.) was studied by Thompson *et al.*, (2012) and hypothesized that changes in the phosphorylation states of mTOR signaling network is involved in the reduction of cancer burden by dietary bean.

Fang *et al.*, (2011) described the purification and characterization of a new *Phaseolus vulgaris lectin* (polygalacturonic acid-specific lectin (termed BTKL)) that exhibited selective toxicity to human hepatoma Hep G2 cells and it lacked significant toxicity on normal liver WRL 68 cells.

An antifungal peptide with a defensin-like sequence and exhibiting a molecular mass of 7.3kDa was purified from dried seeds *of Phaseolus vulgaris* 'Cloud Bean', which exhibited antifungal activity against *Mycosphaerella arachidicola* and *Fusarium oxysporum* in a dose dependent manner. Proliferation of L1210 mouse leukemia cells and MBL2 lymphoma cells was also inhibited by this antifungal peptide (Wu *et al.*, 2011). Similarly, a dimeric 64-kDa hemagglutinin isolated with a high yield from dried *Phaseolus vulgaris* cultivar "French bean number 35" seeds suppressed mycelial growth, inhibited proliferation of hepatoma HepG2 cells and breast cancer MCF-7 cells. It had no antiproliferative effect on normal embryonic liver WRL68 cells (Lam and Ng, 2010). All these studies support and validate the health benefit claims of natural diets and also its use in Chinese or Indian systems of Medicine.

References

Ismail, A., Tiong, N., Tan, S., and Azlan, A. 2009 Antioxidant properties of selected non-leafy vegetables, Nutrition & Food Science, (2009), 39(2):176 – 180.

Barrett, M.L., and Udani, J.K. 2011. A proprietary alpha-amylase inhibitor from white bean (*Phaseolus vulgaris*): a review of clinical studies on weight loss and glycemic control. Nutr J., (2011), 17:10:24.

Celleno, L., Tolaini, M.V., D'Amore, A., Perricone, N.V., and Preuss, H.G. 2007. A Dietary supplement containing standardized *Phaseolus vulgaris* extract influences body composition of overweight men and women. Int J Med Sci. (2007), 4(1):45-52.

Chan, Y.S., Wong, J.H., Fang, E.F., Pan, W., and Ng, T.B. 2012. Isolation of a glucosamine binding leguminous lectin with mitogenic activity towards splenocytes and anti-proliferative activity towards tumor cells. PloS One., (2012), 7(6):e38961.

Costa, N.M., Walker, A.F., and Low, A.G. 1993. The effect of graded inclusion of baked beans (*Phaseolus vulgaris*) on plasma and liver lipids in hypercholesterolaemic pigs given a Western-type diet. Br J Nutr., (1993), 70(2):515-24.

Duke, J.A. 1981. Handbook of Legumes of World Economic Importance. USDA, Bellsville, Maryland. Plenum, New York and London.

Fang, E.F., Pan, W.L., Wong, J.H., Chan, Y.S., Ye, X.J., and Ng, TB. 2011. A new *Phaseolus vulgaris* lectin induces selective toxicity on human liver carcinoma Hep G2 cells. Arch Toxicol., (2011), 85(12):1551-63. .

Helmstädter, A. 2010.Beans and diabetes: *Phaseolus vulgaris* preparations as antihyperglycemic agents. J Med Food. 2010 Apr;13(2):251-4.

Lam SK, Ng TB. 2010. Isolation and characterization of a French bean hemagglutinin with antitumor, antifungal, and anti-HIV-1 reverse transcriptase activities and an exceptionally high yield. Phytomedicine., (2010), 17(6):457-62.

Obiro, W.C., Zhang, T., and Jiang, B. 2008. The nutraceutical role of the *Phaseolus vulgaris* alpha-amylase inhibitor. Br J Nutr., (2008), 100(1):1-12.

Pari, L., and Venkateswaran, S. 2004. Protective role of *Phaseolus vulgaris* on changes in the fatty acid composition in experimental diabetes, (2004), 7(2):204-9.

Preuss, H. G. 2009. Bean amylase inhibitor and other carbohydrate absorption blockers: effects on diabesity and general health. J Am Coll Nutr., (2009), 28(3):266-76.

Rosa, C.O., Costa, N.M., Leal, P.F., and Oliveira, T.T. 1998. The cholesterol-lowering effect of black beans (*Phaseolus vulgaris*, L.) without hulls in hypercholesterolemic rats. Arch Latinoam Nutr., (1998), 48(4):299-305.

Thompson, M.D., Mensack, M.M., Jiang, W., Zhu, Z., Lewis, M.R., McGinley, J.N., Brick, M.A., and Thompson, H.J. 2011.Cell signaling pathways associated with a reduction in mammary cancer burden by dietary common bean (*Phaseolus vulgaris* L.). Carcinogenesis., (2012), 33(1):226 -32.

Venkateswaran, S., Pari, L., and Saravanan, G. 2002. Effect of *Phaseolus vulgaris* on circulatory antioxidants and lipids in rats with streptozotocin-induced diabetes. J Med Food., (2002), 5(2):97-103.

Wu, X., Sun, J., Zhang, G., Wang, H., and Ng, T.B. 2011. An antifungal defensin from *Phaseolus vulgaris* cv. 'Cloud Bean'. Phytomedicine., (2011), 18(2-3):104-9.

8.3. Cowpea (*Vigna unguiculata* L. or *Vigna sinensis* L.)

Cowpea is in cultivation in India and other tropics of world from very ancient times. It is grown for its long green pods (as vegetable), seeds (as pulse) and foliage as vegetable and fodder. It is used as food at both the green shell and dry stage. In Hindi it is called 'Lobia' or 'Chowli'. The Cowpea belongs to the family Leguminosae and sub-family Fabaceae. Vigna is a pan-tropical genus of about 170 species. Three cultivated sub-species have been identified under the species *Vigna unguiculata* L.viz. *V.unguiculata* ssp. *Unguiculata* (dual purpose type), *V. unguiculata* ssp. *Cylindrica* (grain type) and *V. unguiculata* var. *sesquipedalis* (vegetable type).

Cultivars/Varieties: Many promising varieties have been released by institutes and universities. The notable ones are Arka Grima, IIHR-16, Arka Suman, Arka Samrudhi, Pusa Komal, Pusa Phalguni, Pusa Barsati, Pusa Dofasli, Pusa Rituraj, Vyjayanthi, Lola, Bhagylakshmi, Kairali, Varun, Co-2, Vamban, Sel-2-1, Sel-263, Bidhan Barati-1 and Bidhan Barati-2.

Uses

Cowpea grown to maturity can be used as a feed (grazed or harvested for fodder), or its pods can be harvested and eaten as a vegetable. The beans are nutritious and provide complementary proteins to cereals. Some people eat both the fresh pods and leaves, and the dried seeds are popular ingredients in a variety of dishes. Cowpea is a nutritive vegetable which supplies good quality protein, calcium, phosphorus, iron, carotene, thiamine and riboflavin.

Nutritional Components of Cowpea pods

Constituents	Quantity (per 100 g)	Constituents	Quantity (per 100 g)
Moisture	85.3 g	Potassium	242 mg
Carbohydrates	8.10 g	Calcium	72 mg
Proteins	3.50 g	Magnesium	60 mg
Fats	0.20 g	Carotene	564 µg
Energy	48 Kcal	Thiamine	0.07 mg
Dietary Fibre	2.0 g	Riboflavin	0.09 mg
Iron	2.5 mg	Niacin	0.9 mg
Phosphorus	59 mg	Vitamin C	14 mg

Gopalan, *et al.*, (1985)

Phytochemicals

High levels of Vitamin C besides other vitamins present include folate, thiamine and niacin. The minerals magnesium, phosphorus, potassium, calcium, and iron are also present in cowpea in considerable quantities. The phytochemicals identified in cowpea are all carotenoids. However, it is likely that they contain similar compounds to green beans, including flavonoids and chlorophyll. Cowpea contain fibre and are low in calories. Moderate to good antioxidant activity has been found (Marathe *et al.*, 2011), which can provide a number of health benefits. Sprouting increased myo-inositol and glucose content and reduced raffinose which is an inhibitor found in cowpeas (Ribeiro *et al.*, 2011). A protein designated **'unguilin'** was isolated from seeds of the black-eyed pea (*Vigna unguiculata*). It possesses a molecular weight of 18 kDa and an N-terminal sequence resembling that of cyclophilins and the cyclophilin-like antifungal protein from mung beans (Ye and Ng, 2001). Ye *et al.*, (2000) isolated structurally dissimilar proteins with antiviral and antifungal potency from cowpea (*Vigna unguiculata*) seeds.

Light brown, red and black cowpea varieties are rich in antioxidants. The traditional practices of soaking and cooking decreased the antinational factors and improved the protein digestibility of cowpea seeds, while, they have no effect on the biological value and net protein digestibility. To further improve the nutritional quality of cowpea seeds germination followed by soaking and cooking should be used.

Health Benefits of consuming cowpea pods

Cowpeas are an important source of protein in developing countries, especially in West Africa where it is eaten in a variety of ways (Dovlo *et al.*, 1976). Like other legumes cowpeas contribute to the level of dietary protein in starchy tuber-based diets through their relatively high protein content and to the quality of dietary protein by forming complementary mixtures with staple cereals.

- Cowpeas are a source of good protein (23-32%) and dietary fiber thereby helping in weight management, cholesterol reduction, diabetes, cancer, *etc.*
- They have low glycemic index, hence helps in regulating blood sugar levels in diabetics.
- Cowpeas are rich source of lysine and tryptophan, essential for immunity development and mood elevation.
- Cowpeas contain minerals like K, Mg, Ca, P and Fe which play an important role in body fluid maitenance, cardiac function, osteoporosis and blood formation.
- Cowpea protein isolates are known to lower plasma cholesterol due to their specific properties.
- Cowpea contains a few protein inhibitors which are responsible for its anti-proliferative activity and therefore control certain types of cancers.

Research shows that regular consumption of dry beans and other legumes may reduce serum cholesterol, improve diabetic condition, and provide metabolic benefits that aid in weight control (Winham and Hutchins 2007; Anderson *et al.*, 1999), as well as reduce the risk for coronary heart disease (Bazzano *et al.*, 2001; Winham *et al.*, 2007), and cancer (Lanza *et al.*, 2006).

Rotimi *et al.*, (2010) found that while the legumes including cowpea have beneficial effects on reduction of hyperglycemia and strengthening the antioxidant status of the diabetic animals, its increased kidney uric acid concentration is of concern. The study of Joanitti *et al.*, (2010) confirms the anticarcinogenic (breast cancer) potential of Bowman-Birk protease inhibitors found in cowpea.

Enzymatic proteolysis of food proteins is used to produce peptide fractions which has potential to act as physiological modulators. Sengura *et al.*, (2010) investigated the angiotensin-I converting enzyme (ACE-I) inhibitory and antioxidant activities for hydrolysates produced by hydrolyzing *Vigna unguiculata* protein extract as well as compared it with ultrafiltered peptide fractions from these hydrolysates, and postulated that *V. unguiculata* protein hydrolysates and their corresponding ultrafiltered peptide fractions might be utilized for physiologically functional foods with antihypertensive and antioxidant activities.

Plant defensins are small cysteine-rich proteins commonly synthesized in plants, encoded by large multigene families. Most plant defensins that have been characterized to date show potent antifungal and/or bactericidal activities. Different protease inhibitors including Bowman-Birk type (BBI) have been reported from the seeds of *Vigna unguiculata*. The nutritional and physiological effects of raw cowpea (*Vigna ioavailabl* (L) Walp.) seed meal, protein isolate (globulins), or starch on the metabolism of young growing rats were investigated by Olivera *et al.*, (2003). They found that proportional weights of the small intestine and pancreas were increased by raw cowpea seed meal diets, and serum cholesterol levels were slightly reduced. They suggested that these effects were primarily due to the combined actions of globulins, resistant starches, protease inhibitors, and possibly fiber and non-starch polysaccharides on intestinal and systemic metabolism.

'Unguilin' (a protein isolated from cowpea) exerted an antifungal effect toward fungi including *Coprinus comatus, Mycosphaerella arachidicola, and Botrytis ioavail*. In addition, 'unguilin' was capable of inhibiting human immunodeficiency virus-1 reverse transcriptase and the glycohydrolases a- and beta-glucosidases which are involved in HIV infection (Ye & Ng, 2001).

References

Abdelatief, S., and El-Jasser, H. 2010. Chemical and Biological Properties of LocalCowpea Seed Protein Grown in Gizan Region. International Journal of Agricultural and Biological Sciences 1: (2):88-94.

Anderson, J.W., Smith, B.M., and Washnock, C.S. 1999. Cardiovascular and renal benefits of dry bean and soybean intake. *Am. J. Clin Nutr.*, 1999, 70 (3 SUPPL.), 464S-474S.

Bazzano, L.A., He, J., Ogden, L.G., Loria, C., Vupputuri, S., Myers, L., Whelton, P.K. 2001. Legume consumption and risk of coronary heart disease in US men and women: NHANES I Epidemiologic Follow-up Study. Archives Internal Med., 2001, 161, 2573-2578.

http://cowpea.wordpress.com/health-benefits/

Joanitti, G.A., Azevedo, R.B., and Freitas, S.M. 2010. Apoptosis and lysosome membrane permeabilization induction on breast cancer cells by an anticarcinogenic Bowman-Birk protease inhibitor from *Vigna unguiculata* seeds. Cancer Lett., (2010), 293(1):73-81.

Lanza, E., Hartman, T.J., Albert, P.S., Shields, R., Slattery, M., Caan, B., Paskett, and Schatzkin, A. 2006. High dry bean intake and reduced risk of advanced colorectal adenoma recurrence among participants in the polyp prevention trial. J. Nutr., (2006), 136, 1896-1903.

Marathe, S.A., Rajalakshmi, V., Jamdar, S.N., and Sharma, A. 2011. Comparative study on antioxidant activity of different varieties of commonly consumed legumes in India. Food Chem Toxicol., (20110, 49(9):2005-12.

Olivera, L., Canul, R.R., Pereira-Pacheco, F., Cockburn, J, Soldani, F., McKenzie, N.H., Duncan, M., Olvera-Novoa, M.A., and Grant, G. 2003. Nutritional and physiological responses of young growing rats to diets containing raw cowpea seed meal, protein isolate (globulins), or starch. J Agric Food Chem., (2003), 51(1):319-25.

Rao, K.N., and Suresh, C.G. 2007. Bowman-Birk protease inhibitor from the seeds of *Vigna unguiculata* forms a highly stable dimeric structure. Biochim Biophys Acta., (2007), 1774(10):1264-73.

Ribeiro, E.S., Centeno, D.C., Figueiredo, R.C., Fernandes, K.V., Xavier-Filho, J., and Oliveira, A.E. 2011. Free cyclitol, soluble carbohydrate and protein contents in *Vigna unguiculata* and *Phaseolus vulgaris* bean sprouts. J Agric Food Chem., (2011), 59(8):4273-8.

Rotimi, S.O., Olayiwola, I., Ademuyiwa, O., and Adamson, I. 2010. Inability of legumes to reverse diabetic-induced nephropathy in rats despite improvement in blood glucose and antioxidant status. J Med Food., (2010), 13(1):163-9.

Segura, C.M.R., Chel, G.L.A., and Betancur, A.D.A. 2010.Angiotensin-I converting enzyme inhibitory and antioxidant activities of peptide fractions extracted by ultrafiltration of cowpea *Vigna unguiculata* hydrolysates. J Sci Food Agric., (2010), 90(14):2512-8.

Wihman, M. D., Hutchins, A.M., and Johnston, C.S. 2007. Pinto bean consumption reduces biomarkers for heart disease risk. J. Am. College Nutr., (2007), 26:243-249.

Ye, X.Y, and Ng, T.B. 2001. Isolation of unguilin, a cyclophilin-like protein with anti-mitogenic, antiviral, and antifungal activities, from black-eyed pea. J Protein Chem., (2001), 20(5):353-9. Doi.org/10.1023/A:1012272518778.

Ye, X.Y., Wang, H.X., and Ng, T.B. 2000. Structurally dissimilar proteins with antiviral and antifungal potency from cowpea (*Vigna unguiculata*) seeds. Life Sci. (2000), 67(26):3199-207.

9. Bulbous Crops

9.1. Onion (*Allium cepa* L.)

The onion (*Allium cepa*), which is also known as the bulb onion, or common onion is the most widely cultivated species of the genus Allium. It has been used as food since time immemorial and is an important vegetable all over the world. The genus Allium also contains a number of other species variously referred to as onions and cultivated for food, such as the Japanese bunching onion (*A. fistulosum*), Egyptian onion (*A. proliferum*), and Canada onion (*A. canadense*). The vast majority of cultivars of *A. cepa* belong to the "common onion group" (*A. cepa var. cepa*) and are usually referred to simply as "onions". The Aggregatum Group of cultivars (*A. cepa var. aggregatum*) includes both shallots and potato onions. Onion is suggested to have been originated in Asia, particularly Pakistan and Iran.

China is the largest onion producer in the world followed by India. Largest area under onion is in India followed by China. The major onion producing state are Maharashtra, Karnataka, Madhya Pradesh, Rajasthan, Gujarat, Bihar, Andhra Pradesh, Tamil Nadu and Orissa.

Cultivars/Varieites: Onion varieties differ in size, colour of skin, pungency and maturation. The most popular varieties under cultivation are Agrifound Dark Red, Agrifound Light Red, Agrifound Red, Agrifound Rose, Arka Bindu, Arka Kalyan, Arka Niketan, Arka Pragati, Baswant 780, Brown Spanish, Co-1, Co-2 Co- 3, Co-4, Early Grano, Hisar II, Kalyanpur Red Round, MDU 1 N-2-4-1, N-53, N-257-9-1, Punjab 48 (S 48), Pusa Madhavi, Punjab Red Round, Punjab Selection, Pusa Ratnar, Pusa Red, Pusa White Flat, Pusa White Round, Udaipur 101, 102 & 103.

Uses

Onions are used for flavouring or seasoning of food both at mature and immature bulb stages, besides being used as salad and pickle. To some extent it is used by processing industry for flakes or powders. Onions are often chopped and used as an ingredient in various dishes. Onions are also used as a thickening agent for curries providing a bulk of the base or gravy. Onions pickled in vinegar are eaten as a snack. These are often served as a side serving in fish and chicken fries. Fresh onion has a pungent, persistent, even irritating taste, but when sauteed, onion becomes sweet and much less pungent.

Onions contain a lot of phytochemicals like phenols and flavonoids besides vitamins and minerals.

Nutritional Components of Onion

Constituents	Quantity (per 100 g)	Constituents	Quantity (per 100 g)
Moisture	87.11 g	Calcium	23 mg
Carbohydrates	9.34 g	Magnesium	0.129 mg
Proteins	1.10 g	Zinc	0.17 mg
Fats	0.10 g	Thiamine	0.046 mg
Energy	40 Kcal	Riboflavin	0.027 mg
Dietary Fibre	1.70 g	Niacin	0.116 mg
Iron	0.21 mg	Vitamin C	7.4 mg
Phosphorus	29 mg	Vitamin E	0.02 mg
Potassium	146 mg	Vitamin K	0.4 µg

Gopalan et al., (1985)

Phytochemicals

Onions are a very good source of vitamin C, B6, biotin, chromium, calcium and dietary fibre. In addition, they contain good amounts of folic acid and vitamin B1 and K. The onion is the richest dietary source of quercitin, a potent antioxidant flavonoid (also in shallots, yellow and red onions only but not in white onions), which is found on and near the skin and is particularly linked to the health benefits of onions.

They also contain specific amino acids called methionine and cystine. Phytochemicals in onions include the organosulfur compounds such as cepaenes and thiosulfinates (Dorsch and Wagner, 1991; Goldman et al., 1996), the large class of flavonoids including quercetin and kaempferol (Dorant et al., 1994), and pigments such as anthocyanins found in red onions. The onions are characterized by their rich content of thiosulfinates, allyl sulfides, sulfoxides, and other odoriferous sulfur compounds. The alk(en)yl cysteine sulfoxides (ACSOs) are primarily responsible for the onion flavor and produce the eye-irritating compounds that induce lacrimation. Sixteen different flavonols consisting of aglycones and glygosylated derivatives of quercitin, isorhamnetin and kaempferol have been identified (Bilyk et al., 1984). Rutin and myricetin are also found in small quantities. Onions are a very rich source of fructo-oligosaccharides. These oligomers stimulate the growth of healthy bifidobacteria and suppress the growth of potentially harmful bacteria in the colon. In addition, they can reduce the risk of tumors developing in the colon. The thiosulfinates exhibit antimicrobial properties.

Health Benefits of Consuming Onions and the Research Findings

Onions have received considerable attention for their health promoting, functional benefits. The organosulfur compounds are believed to possess anti-inflammatory, anti-allergic, anti-microbial, and anti-thrombotic activity by inhibition of cyclooxygenase and lipoxygenase enzymes. Quercetin and kaempferol, the major flavonoids in onions, are found in the flavonol subclass. Quercetin is the major flavonoid of interest in onions. Mechanisms of action include free radical scavenging, chelation of transition metal ions, and inhibition of oxidases such as lipoxygenase (de Groot and Rauen, 1998; Suzuki et al., 1998; Lean et al., 1999).

Anti-Diabetic: Significant amount of research has been carried out on the effect of onion consumption on diabetic conditions. The organosulfur compounds S-methylcysteine sulfoxide (SMCS) and S-allylcysteine sulfoxide (SACS) were linked to amelioration of weight loss, hyperglycemia, low liver protein and glycogen, and other characteristics of diabetes mellitus in rats (Sheela *et al.*, 1995). They found that the use of SMCS and SACS (200mg/kg/day) gave results comparable to treatment with insulin or glibenclamide but without the negative side effect of cholesterol synthesis stimulation. Similarly, Suresh and Srinivasan (1997) found that a 3% onion powder di*et al.*,so reduced hyperglycemia, circulating lipid peroxides, and blood cholesterol (LDL-VLDL exclusively). *In vivo* analysis of the effects of quercetin on human diabetic lymphocytes showed a significant increase in the protection against DNA damage from hydrogen peroxide at the tissue level (*Lean et al.*, 1999).

One of the therapeutic approaches for decreasing postprandial hyperglycemia is to retard absorption of glucose by the inhibition of carbohydrate hydrolyzing enzymes, α-amylase and α-glucosidases, in the digestive organs. Extract of onion skin was found to improve exaggerated postprandial spikes in blood glucose and glucose homeostasis since it inhibits intestinal sucrase and thus delays carbohydrate absorption (Kim *et al.*, 2011). Diabetic Neuropathy (DN) is a major microvascular complication of uncontrolled diabetes. This may result from increased oxidative stress that accompanies diabetes. Plants with antioxidant action play an important role in management of diabetes and its complications. Bhanot and Richashri (2010) showed that the extract of the outer scale of onion and the edible portion of both onion and garlic provided significant protection in DN in both preventive and curative groups. The methanol extract of the outer scales of onion has shown a most significant effect which may be due to the presence of higher quantities of phenolic compounds.

Anti-Cancerous: The inhibitory effects of onion consumption on human carcinoma have been widely researched. Epidemiological data both support (Gao *et al.*, 1999; Hu *et al.*, 1999) and refute (Dorant *et al.*, 1995) the concept that higher intake of onions is positively related to lower risk for carcinoma. In a review on the effects of quercetin, Hertog and Katan (1998) noted that persons in the highest consumption category versus the lowest had a 50% reduced risk of cancers of the stomach and alimentary and respiratory tracts. Organosulfur compounds such as diallyl disulfide (DDS), S-allylcysteine (SAC), and S-methylcysteine (SMC) have been shown to inhibit colon and renal carcinogenesis (Hatono *et al.*, 1996; Fukushima *et al.*, 1997). Mechanisms of protection ranged from induced cancer cell apoptosis (Richter *et al.*, 1999) and gene transcription inhibition (Miodini *et al.*, 1999) to protection against UV-induced immunosuppression (Steerenberg *et al.*, 1998).

Anti-Hyperlipidemia: Limited information from human studies indicates that dietary quercetin supplementation influences blood lipid profiles, glycemic response, and inflammatory status, collectively termed cardiometabolic risks. Lee *et al.*, (2011) hypothesized that quercetin-rich supplementation, derived from onion peel extract, improves cardiometabolic risk components in healthy male smokers and found that its improved blood lipid profiles, glucose, and blood pressure, suggesting a beneficial role for quercetin as a preventive measure against cardiovascular risk.

Ige et al., (2009) carried out a research work, targeted at knowing whether or not oral administration of the AcE will prevent cadmium's adverse effects of on renal clearance as a representative of renal functions. Their findings lead to the conclusion that cadmium (Cd) exposure causes renal dysfunction, but oral administration of onion extract could prevent it.

Anti-Osteoporotic: Muhlbauer and Li (1999) demonstrated through animal studies that with onion intake by rats there was considerable increase in bone mass, bone thickness, and bone mineral density. Onions inhibited bone resorption by 20% when consumed at a rate of 1g per day per kg of body weight. This was slightly higher than the rate of bone resorption obtained from the calcitonin that is typically used to treat postmenopausal osteoporosis. These findings suggest that onion intake may be a useful dietary approach to improving bone health.

Cardiovascular health: Inhibition of LDL oxidation and platelet aggregation were proposed as mechanisms of benefit against cardiovascular disease (Janssen et al., 1998). Quercetin exerts its beneficial effects on cardiovascular health by antioxidant and anti-inflammatory activities (Anonymous, 1998; Kuhlmann et al., 1998). Adenosine and paraffinic polysulfides (PPS) are compounds isolated from onions with purported antiplatelet effects (Makheja and Bailey, 1990; Augusti, 1996; Yin and Cheng, 1998). Researchers found that presence of quercetin significantly reduced LDL oxidation in vitro from various oxidases including 15-lipoxygenase, copper-ion, UV light, and linoleic acid hydroperoxide (Nègre-Salvayre and Salvayre, 1992; da Silva et al., 1998; Aviram et al., 1999; Kaneko and Baba, 1999). Besides the direct antioxidant effect, quercetin also inhibited consumption of alpha-tocopherol (Hertog and Katan, 1998; da Silva et al., 1998; Kaneko and Baba, 1999) and protected human serum paraxonase (PON 1) activities (Aviram et al., 1999).

Immunosuppression: Quercetin's anti-inflammatory effect on prostaglandins, leukotrienes, histamine release and subsequent antiasthmatic activity has been investigated by several researchers (Wagner et al., 1990; Anonymous, 1998). Quercetin was shown to suppress both immune and non-immune injury responses. Thiosulfinates and capaenes responsible for the anti-inflammatory activities in onions also cause inhibition of the immune response (Dorsch et al., 1990; Chisty et al., 1996), thus serving the purpose of immunosuppression.

HIV Control: Viral protein R (vpr) has been shown to control the rate of replication of HIV-1 (Cohen et al., 1990). Therefore, suppression of this gene is a probable target for inhibition of the development of AIDS. Westervelt et al., (1992) showed that disruption of the functionality of the vpr gene attenuated HIV-1 replication. It was also shown that quercetin may diminish virus replication by inhibiting vpr function (Shimura et al., 1999). At 10µM dosage, quercetin provided 92% inhibition of vpr-induced cell cycle abnormality.

Antimicrobial activity: Although thought to be less active than garlic, onions have been shown to possess antibacterial and antifungal properties (Hughes and Lawson, 1991; Augusti, 1996). Onion oil has been shown to be highly effective against gram positive bacteria, dermatophytic fungi, and growth and aflatoxin production of Aspergillus fungi genera (Zohri *et al.*, 1995).

References

Anonymous. 1998. Quercetin. Alternative Medicine Review. 3(2): 140-143.

Augusti, K. 1996. Therapeutic values of onion and garlic. Indian Journal of Experimental Biology. 34: 634-640.

Aviram, M., Rosenblat, M., Billecke, S., Erogul, J., Sorenson, R., Bisgaier, C., Newton, R., and La Du, B. 1999. Human serum paraoxonase is inactivated by oxidized low density ioavailabl and preserved by antioxidants. Free Radical Biology and Medicine. 26(7/8): 892-904.

Bhanot, A., and Richashri, A. 2010. Comparative profile of methanol extracts of *Allium cepa* and *Allium sativum* in diabetic neuropathy in mice. Pharmacognosy Res., (2010), 2(6): 374–384.

Bilyk, A., Cooper, P.L., and Saper, G.M. 1984. Varietal differences in distribution of quercitin and kaempferol in onion (*Allium cepa* L.) tissue. J. Agric Food Chem., 32:274-281.

Chisty, M., Quddus, R., Islam, B., and Khan, B. 1996. Effect of onion extract on immune response in rabbits. Bangladesh Med. Res. Counc. Bull. 22: 81-85.

Cohen, E., Terwilliger, E., Jalinoos, Y., Proulx, J., Sodroski, J., and Hasiltine, W. 1990. Identification of HIV-1 vpr product and function. Journal of Acquired Immune Deficiency Syndromes. 3: 11-18.

da Silva, E., Tsushida, T., and Terao, J. 1998. Inhibition of mammalian 15-lipoxygenase-dependent lipid peroxide in low-density lipoprotein by quercetin and quercetin monoglucosides. Archives of Biochemistry and Biophysics. 349(2): 313-320.

de Groot, H., and Rauen, U. 1998. Tissue injury by reactive oxygen species and the protective effects of flavonoids. Fundam. Clin. Pharmacol., 12: 249-255.

Dorant, E., Van Den Breandt, P., and Goldbohm, A.1995. Allium vegetable consumption, garlic supplement intake, and female breast carcinoma incidence. Breast Cancer Research and Treatment, 33: 163-170.

Dorant, E., Van Din Brandt, P., and Goldbohm, R. 1994. A prospective cohort study on Allium vegetable consumption, garlic supplement use, and the risk of lung carcinoma in the Netherlands. Cancer Research, 54: 6148-6153.

Dorsch, W., and Wagner, H. 1991. New antiasthmatic drugs from traditional medicine? Int Arch Allergy Appl Immunol., 94: 262-265.

Dorsch, W., Schneider, E., Bayer, T., Breu, W., and Wagner, H. 1990. Anti-inflammatory effects of onions: inhibition of chemotaxis of human polymorphonuclear leukocytes by thiosulfinates and cepaenes. Int. Arch. Allergy Appl. Immunol., 92: 39-42.

Fukushima, S., Takada, N., Hori, T., and Wanibuchi, H. 1997. Cancer prevention by organosulfur compounds from garlic and onion. Journal of Cellular Biochemistry Supplement, 27: 100-105.

Gao, C., Takezaki, T., Ding, J., Li, M., and Tajima, K.1999. Protective effect of Allium vegetables against both esophageal and stomach cancer: A simultaneous casereferent study of a high-epidemic area in Jiangsu province, China. Jpn. J. Cancer Res., 90: 614-621.

Goldman, I., Kopelberg, M., Devaene, J., and Schwartz, B.1996. Antiplatelet activity in onion is sulfur dependent. Thrombosis and Haemostasis, 76(3):450-452.

Hatono, S., Jimenez, A., and Wargovich, M. 1996. Chemopreventive effect of S-allylcysteine and the relationship to the detoxification enzyme glutathione S-transferase. Carcinogenesis, 17(5): 1041-1044.

Hertog, M., and Katan, M. 1998. Quercetin in foods, cardiovascular disease, and cancer. Ch. 20, In: Flavonoids in Health and Disease, p. 447-467.

http://en.wikipedia.org/wiki/Onion

Hu, J., Vecchia, C., Negri, E., Chatenoud, L., Bosetti, C.,Jia, X., Liu, R., Huang, G., Bi, D., and Wang, C. 1999. Diet and brain cancer in adults: a case-control study in northeast China. Int. J. Cancer, 81: 20-23.

Hughes, B., and Lawson, L. 1991. Antimicrobial effects of *Allium sativum* L. (garlic), *Allium ampeloprasum* L. (elephant garlic), and *Allium cepa* L. (onion), garlic compounds and commercial garlic supplement products. Phytotherapy Research, 5: 154-158.

Ige, S. F., Salawu, E. O., Olaleye, S. B., Adeeyo, O. A., Badmus, J., and Adeleke, A. A. 2009. Onion (*Allium cepa*) extract prevents cadmium induced renal dysfunction. Indian J Nephrol., (2009), 19(4): 140–144.

Janssen, K., Mensink, R., Cox, F., Harryvan, J., Hovenior, R., Hollman, P., and Katan, M. 1998. Effects of the flavonoids quercetin and apigenin on hemostasis in healthy volunteers: results from an *in vitro* and a dietary supplement study. Am J Clin Nutr., 2: 255-262.

Kaneko, T., and Baba, N. 1999. Protective effect of flavonoids on endothelial cells against linoleic acid hydroperoxide-induced toxicity. Biosci. Biotechnol. Biochem., 63(2): 323-328.

Kuhlmann, M., Burkhardt, G., Horsch, E., Wagner, M., and Kohler, H. 1998. Inhibition of oxidant-induced lipid peroxidation in cultured renal tublar epithelial cells by quercetin. Free Rad. Res., 29: 451-460.

Lean, M., Noroozi, M., Kelly, I., Burns, J., Talwar, D., Satter, N., and Crozier, A. 1999. Dietary flavonoids protect diabetic human lymphocytes against oxidant damage to DNA. Diabetes, 48: 176-181.

Lee, Kyung-Hea., Park, Eunj., Lee, Hye-Jin., Kim, Myeong-Ok., Cha, Yong-Jun., Kim, Jung-Mi., Lee, Hyeran., and Shin, Min-Jeong. 2011. Effects of daily quercetin-rich supplementation on cardiometabolic risks in male smokers Nutrition Research and Practice (Nutr Res Pract)., (2011), 5(1):28-33.

Makheja, A., and Baily, J. 1990. Antiplatelet constituents of garlic and onion. Agents and Actions, 29: 360-363.

Miodini, P., Gioravanti, L., Di Fronzo, G., and Cappelletti, V. 1999. The two phyto-oestogens genistein and quercetin exert different effects on oestrogen receptor function. British J. Cancer, 80(8): 1150-1155.

Mulbauer, R.C., and Li, F. 1999. Effect of vegetables on bone metabolism. Nature, 401:343-344.

Nègre-Salvayre, A., and Salvayre, R. 1992. Quercetin prevents the cytotoxicity of oxidized LDL of lymphoid cell lines. Free Radical Biology and Medicine, 12: 101-106.

Richter, M., Ebermann, E., and Marian, B. 1999. Quercetin-induced apoptosis in colorectal tumor cells: possible role of EGF receptor signaling. Nutrition and Cancer, 34: 88-99.

Sheela, C., Kumud, K., and Augusti, K. 1995. Anti-diabetic effects of onion and garlic sulfoxideamino acids in rats. Planta Med., 61: 356-357.

Shimura, M., Zhou, Y., Asada, Y., Yoshikawa, T., Hatake, K., Takaku, F., and Ishizaka, Y. 1999. Inhibition of Vpr-induced cell cycle abnomality by quercetin: A novel strategy for searching compounds targeting Vpr. Biochemical and Biophysics Research Communications, 261: 308-316.

Steerenberg, P.A., Garssen, J., Dortant, P., Hollman, P.C., Alink, G.M., Dekker, M., Bueno-de-Mesquita, H.B., and Van Loveren, H. 1998. Protection of UV-induced suppression of skin contact hypersensitivity: A common feature of flavonoids after oral administration? Photochemistry and Photobiology, 67(4): 456-461.

Babu, Suresh, P., and Srinivasan, K. 1997. Influence of dietary capaicin and onion an the metabolic abnormalitiesassociated with streptozotocin induced diabetes mellitus. Molecular and Cellular Biochemistry, 175: 49-57.

Suzuki, Y., Masashi, I., Segami, T., and Ito, M. 1998. Anti-ulcer effects of antioxidant, quecetin, a-tocopherol, nifedipine and tetracycline in rats. Jpn. J. Pharmacol., 78: 435-441.

Wagner, H., Dorsch, H., Bayer, T., Breu, W., and Willer, F. 1990. Antiasthmatic effects of onions: inhibition of 5-lipoxygenase and cyclooxygenase in vitro by hiosulfinates and "Cepaenes." Prostaglandins Leukotrienes and Essential Fatty Acids, 39: 59-62.

Westervlt, P., Henkel, T., Trowbridge, D., Orenstein, J., Heuser, J., Gendlman, H., and Ratner, L. 1992. Dual regulation of silent and productive infection in monocytes by distinct human immunodeficiency virus type 1 determinants. Journal of Virology, 66(6): 3925-3931.

Yin, M., and Cheng, W. 1998. Antioxidant activity of several Allium members. J. Agric. Food Chem., 46: 4097-4101.

Zohri, A., Abdel-Gawad, K., and Saber, S. 1995. Antibacterial, antidermatophytic and antitioxigenic activities of onion (Allium cepa L.) oil. Microbiol. Res., 150: 167-172.

9.2. Garlic (*Allium sativum* L.)

Allium sativum, commonly known as garlic, is a species in the onion genus, Allium. Its close relatives include the onion, shallot, leek, and chive. It is native to central Asia and has long been a staple in the Mediterranean region, as well as in Asia, Africa, and Europe. It was known to ancient Egyptions and has been used for both culinary and medicinal purposes. The mention of use of garlic for many conditions, including parasites, respiratory problems, poor digestion, and low energy has been made by even Hippocrates.

Among the Allium species, garlic follows onion in area and production in India. Madhya Pradesh, Gujarat, Uttar Pradesh, Maharashtra, Karnataka and Jammu and Kashmir are the major producers of garlic.

Cultivars/Varieties: Garlic varieties differ in size, colour and pungency of cloves. The popular varieties are Agrifound White (G-41), Yamuna Safed (G-1), Yamuna Safed-2 (G-50), Yamuna Safed-3 (G-282), Agrifound Parvati (G-323), Ooty-1 and Pant Lohit. Loca types are also grown viz., Fawari & Rajalle Gaddi in Karnataka, Kanthalloor Local in Kerala and Tabiti and Jamnagar in Gujarat.

Nutritive Value of Garlic

Nutritional Components of Garlic			
Constituents	Quantity (per 100 g)	Constituents	Quantity (per 100 g)
Moisture		Calcium	181 mg
Carbohydrates	33.6g	Magnesium	25 mg
Proteins	6.39 g	Zinc	1.16 mg
Fats	0.5 g	Thiamine	0.20 mg
Energy	149 Kcal	Riboflavin	0.11 mg
Dietary Fibre	2.10 g	Niacin	0.7 mg
Iron	1.70 mg	Vitamin C	31.2 mg
Phosphorus	153 mg	Vitamin B6	1.24 mg
Potassium	401 mg	Pantothenic acid	0.596 mg

Gopalan *et al.*, (1985)

Garlic promotes the well-being of the heart and immune systems with antioxidant properties and helps maintain healthy blood circulation. One of garlic's most potent health benefits includes its ability to enhance the body's immunity.

Phytochemicals

The sulfur compound allicin, produced by crushing or chewing fresh garlic, produces other sulfur compounds like ajoene, allyl polysulfides, and vinyldithiins. Aged garlic lacks allicin but may have some activity due to the presence of S-allylcysteine. The following sulfur-containing constituents are abundantly found in garlic that help lower the risk of oxidative stress: Alliin, allicin, allixin, allyl polysulfides (APS -Allyl polysulfides is a general term that refers to a variety of compounds), diallyl sulfide (DAS), diallyl disulfude (DADS), diallyl trisulfide (DATS),N-acetylcysteine (NAC),

N-acetyl-S-allylcysteine (NASC), S-allylcysteine (SAC), S-allylmercaptocysteine (SAMC), S-ethylcysteine (SEC), S-methylcysteine (SMC), S-propylcysteine (SPC), 1, 2-vinyldithiin (1,2-DT) and thiacremonone.

Health Benefits of Consuming Garlic and the Research Findings

Garlic products have become more popular in the last couple of decades. Market research conducted in United States (1998) showed that garlic products were the most popular of all 91 dietary supplements (Wyngate, 1998). Dozens of brands on store shelves can be classified into four groups: garlic oil, garlic oil macerate, garlic powder and aged garlic extract.

- Aged Garlic Extract (AGE) has been found to have antibacterial, antiviral, and antifungal activity.
- Other beneficial properties include preventing and fighting the common cold. In Chinese Medicine garlic is used for hoarseness and coughs. Garlic was used as an antiseptic to prevent gangrene during World War I and World War II.
- Garlic cloves are used as a remedy for infections (especially chest problems)., digestive disorders, and fungal infections such as thrush. Garlic can be used as a disinfectant because of its bacteriostatic and bacteriocidal properties.

Antioxidant Activity: Several reports have shown that AGE and SAC inhibit the oxidative damage implicated in aging and a variety of diseases. Derived of these works, different antioxidant mechanisms have been attributed to these compounds and confirmed. Most of the protective properties of garlic and its products are brought out by the antioxidant capacity of different phytochemicals of garlic mentioned above. However, some specific mechanisms also operate in certain pathways of prevention or amelioration.

Cardioprotective Property: Garlic is also claimed to help prevent heart disease, including atherosclerosis, high cholesterol, and high blood pressure. Animal studies, and some early research studies in humans, have suggested possible cardiovascular benefits of garlic. A Czech study found garlic supplementation reduced accumulation of cholesterol on the vascular walls of animals. Garlic supplementation significantly reduced aortic plaque deposits of cholesterol-fed rabbits. Studies showed supplementation with garlic extract inhibited vascular calcification in human patients with high blood cholesterol. The known vasodilative effect of garlic is possibly caused by catabolism of garlic-derived polysulphides to hydrogen sulphide in red blood cells (RBCs), a reaction that is dependent on reduced thiols in or on the RBC membrane. Hydrogen sulfide is an endogenous cardioprotective vascular cell-signaling molecule. Bhatti *et al.*, (2008) demonstrated that garlic extract exaggerates the cardio protection offered by ischemic preconditioning. Treatment with garlic extract also protects the myocardium against ischemia reperfusion induced cardiac injury

Controls Hyperlipidemia: Considerable evidence has indicated that allicin is responsible for most of the effects of garlic on serum lipids (Lawson, 2010). Garlic supplements in powder form was compared for tis active prinicple activity and it was found that the powder supplements also have same activity as that of freshly crushed garlic (Lawson and Gardner, 2008). Garlic is best known for its lipid lowering and anti-atherogenic effects. Possible mechanisms of action include inhibition of the hepatic activities of lipogenic and cholesterogenic enzymes that are thought to be the genesis for dyslipidemia, increased excretion of cholesterol and suppression of LDL-oxidation. Quinna *et al.*, (2012) evaluated the anti-dyslipidemic properties of the combination of the artichoke leaves extract, turmeric extract, prickly pear dried leaves (PPL) and garlic extract versus each one alone in two different hyperlipidemic animal models and found that treatment using the combination of artichoke, turmeric, PPL and garlic extract prevents dyslipidemia; partially through inhibiting HMG-CoA reductase.

Often the herbal supplements are taken together with drugs for various kinds of ailments assuming that herbals are safe or no side effects. Matten *et al.*, (2011) studied the effect of co-administration of antiplatelet drug (cilostazol) with garlic extract and found no pharmacodynamic interactions between garlic and cilostazol.

Antidiabetic Property: Regular and prolonged use of therapeutic amounts of aged garlic extracts lower blood homosysteine levels and has been shown to prevent some complications of diabetes mellitus. Oxidative damage by free radicals has been implicated in the pathogenesis of vascular disease in diabetes and hypertension. The total antioxidant status in diabetic and hypertensive rats before and after treatment with garlic (*Allium sativum*) was determined and it was found that the total antioxidant status can be significantly improved by treatment with garlic (Drobiova *et al.*, 2010).

Antibiotic Property: Herbs and spices are very important and useful as therapeutic agent against many pathological infections. Increasing multidrug resistance of pathogens forces to find alternative compounds for treatment of infectious diseases. Garlic cloves are used as a remedy for infections (especially chest problems) in traditional medicine. Gull *et al.*, (2012) demonstrated the inhibitory effect of aqueous, methanolic and ethanolic extracts of garlic on *Escherichia coli, Pseudomonas aeruginosa, Bacillus subtilis, Staphylococcus aureus, Klebsiella pneumoniae, Shigella sonnei, Staphylococcusepidermidis* and *Salmonella typhi*. Thirty strains of mycobacteria, consisting of 17 species, were inhibited by various concentrations of garlic extract incorporated in Middlebrook 7H10 agar (Delaha and Garagusi, 1985). Plasma titers of anti-*Cryptococcus neoformans* activity rose twofold over preinfusion titers and anti-*C. neoformans* activity was detected in four of five cerebrospinal fluid samples of meningitis patients administered intravenously with commercial garlic extract (Davis *et al.*, 1990).

Hepatoprotective Activity: As in case of onion, the garlic is also effective in reducing the deleterious effect of heavy metals on liver. Sharma *et al.*, (2010) studied the effect of garlic extract on lead induced hepatotoxicity in albino rats and suggested that garlic is a phyto-antioxidant that can counteract the deleterious effects of lead nitrate on liver. Similarly, Shaarawy *et al.*, (2009) also found garlic to have hepatoprotective activity in NDEA induced hepatotoxicity studies.

Nephroprotective Activity: Mercury is one of the heavy metals which causes nephrotoxity and several other renal dysfunctions. Simultaneous administration of garlic along with mercuric chloride, was found to produce a pronounced nephroprotective effect against mercuric chloride induced toxicity in rats by restoring the normal levels of biochemical parameters (Abirami and Jagadeewari, 2006).

Neuroprotective Property: Oxidative stress caused by increased accumulation of reactive oxygen species (ROS) in cells has been implicated in the pathophysiology of several neurodegenerative diseases including Alzheimer's disease (AD). Several studies have demonstrated the antioxidant properties of garlic and its different preparations including Aged Garlic Extract (AGE). AGE and S-allyl-cysteines (SAC), a bioactive and bioavailable component in garlic preparations have been shown in a number of *in-vitro* studies to protect neuronal cells against beta-amyloid (A) toxicity and apoptosis. Mathew and Biju (2008) reviewed the neuroprotective effect of garlic and summarized that the broad range of anti-atherogenic, antioxidant and anti-apoptotic protection afforded by garlic may be extended to its neuroprotective action, helping to reduce the risk of dementia, including vascular dementia and AD.

Analgesic Property: Patients with osteoarthritis commonly use complementary and alternative medicines (CAM), either as an adjunct to or in place of conventional analgesics. Vitamin supplementation was the most common CAM reported, followed by celery extract, fish oils, and garlic extracts (Zochling *et al.*, 2004).

Antianemic Property: Sickle cell anemia is one of the most prevalent hereditary disorders with prominent morbidity and mortality. With this disorder oxidative, phenomena play a significant role in its pathophysiology. One of the garlic (*Allium sativum* L.) formulations, aged garlic extract (AGE), has been reported to exert an anti-oxidant effect *in vitro*, Takasu *et al.*, (2002) evaluated the anti-oxidant effect of AGE on sickle red blood cells (RBC) and suggested that there is a significant anti-oxidant activity of AGE on sickle RBC. They opined that AGE may be further evaluated as a potential therapeutic agent to ameliorate complications of sickle cell anemia.

Anti-Cancer Property: Garlic is used to prevent certain types of cancer, including stomach and colon cancers. Published results of epidemiologic case-control studies in China and Italy on gastric carcinoma in relation to diet suggest that consuming garlic may reduce the risk of gastric cancer. Chemical constituents of garlic have been tested for their inhibiting effect on carcinogenesis, using *in vitro* and *in vivo* models. In most experiments' inhibition of tumour growth was established using fresh garlic extract, garlic compounds or synthetically prepared analogs. Dorant *et al.*, (1993) reviewed the strengths and weaknesses of the experiments and discussed the outcomes. Garlic has many health benefits and has been traditionally used worldwide over the centuries. The wealth of scientific literature supports the proposal that garlic and its preparations help in preventing or reducing the risk of cardiovascular ailments, stroke, and cancer. Recently the beneficial effects of garlic and its constituents on neuronal physiology and brain function are beginning to emerge. This review encompasses multiple health effects of garlic and its constituents with references to neuroprotection. Further studies should be carried out to identify specific compounds from garlic that are responsible for most of its biological effects.

References

Abirami, N., and Jagadeeswari, R. 2006. Amelioration of mercuric chloride induced nephrotoxicity and oxidative stress by garlic extract. Anc Sci Life., (2006), 26(1-2):73-7.

Bhatti, R., Singh, K., Ishar, M.P., and Singh, J. 2008. The effect of Allium sativum on ischemic preconditioning and ischemia reperfusion induced cardiac injury. Indian J Pharmacol., (2008), 40(6):261-5.

Davis, L.E., Shen, J.K., and Cai, Y. 1990. Antifungal activity in human cerebrospinal fluid and plasma after intravenous administration of Allium sativum. Antimicrob Agents Chemother., (1990), 34(4) :651-3.

Delaha, E.C., and Garagusi, V.F. 1985. Inhibition of mycobacteria by garlic extract (*Allium sativum*). Antimicrob Agents Chemother., (1985), 27(4):485-6.

Dorant, E., van den Brandt, P.A., Goldbohm, R.A., Hermus, R.J., and Sturmans, F. 1993. Garlic and its significance for the prevention of cancer in humans: a critical view. Br J Cancer., (1993), 67(3):424-9.

Drobiova, H., Thomson, M., Al-Qattan, K., Peltonen-Shalaby, R., Al-Amin, Z, and Ali, M. 2010. Garlic increases antioxidant levels in diabetic and hypertensive rats determined by a modified peroxidase method. Evid Based Complement Alternat Med., (2011), 2011:703049. Doi: 10.1093/ecam/nep011.

Gull, I., Saeed, M., Shaukat, H., Aslam, S.M., Samra, Z.Q., and Athar, A.M. 2012. Inhibitory effect of *Allium sativum* and *Zingiber officinale* extracts on clinically important drug resistant pathogenic bacteria. Ann Clin Microbiol Antimicrob., (2012), 27;11:8.

Larry, L.D., and Gardner, C.D. 2008. Composition, Stability, and Bioavailability of Garlic Products Being Used in a Clinical Trial. J Agric Food Chem., (2005), 53(16): 6254–6261.

Lawson, L.D. 1998. Garlic: a review of its medicinal effects and indicated active compounds. In: Lawson, LD.; Bauer, R., (editors) Phytomedicines of Europe: Chemistry and biological activity. Washington, DC: American Chemical Society; 1998. P. 176-209.

Mateen, A.A., Usharani, P, Rani, Naidu, M.U.R., and Chandrashekar, E. 2011. Pharmacodynamic interaction study of *Allium sativum* (garlic) with cilostazol in patients with type II diabetes mellitus. Indian J. Pharmcol., 43(3): 270-274.

Mathew, B., and Biju, R. 2008. Neuroprotective effects of garlic a review. Libyan J Med., (2008), 3(1):23-33.

Qinna, N.A., Kamona, B.S., Alhussainy, T.M., Taha, H., Badwan, A.A., Matalka, K.Z. 2012. Effects of prickly pear dried leaves, artichoke leaves, turmeric and garlic extracts, and their combinations on preventing dyslipidemia in rats. ISRN Pharmacol., (2012), 2012:167979.

Seo, D.Y., Lee, S.R., Kim, H.K., Baek, Y.H., Kwak, Y.S., Ko, T.H., Kim, N., Rhee, B.D., Ko, K.S., Park, B.J., and Han, J. 2012. Independent beneficial effects of aged garlic extract intake with regular exercise on cardiovascular risk in postmenopausal women. Nutr Res Pract., (2012), 6(3):226-31.

Shaarawy, S.M., Tohamy, A.A., Elgendy, S.M., Elmageed, Z.Y., Bahnasy, A., Mohamed, M.S., Kandil, E., and Matrougui, K. 2009. Protective effects of garlic and silymarin on NDEA-induced rats' hepatotoxicity. Int J Biol Sci., (2009), 5(6):549-57.

Sharma, A., Sharma, V., and Kansal L. 2010. Amelioration of lead-induced hepatotoxicity by Allium sativum extracts in Swiss albino mice. Libyan J Med., (2010), 7:5.

Takasu, J., Uykimpang, R., Sunga, M., Amagase, H., and Niihara, Y. 2002. Aged garlic extract therapy for sickle cell anemia patients. BMC Blood Disord., (2002), 2(1):3.

Wyngate, P. 1998. Phase One study of vitamins, minerals, herbs and supplements research conducted by Hartman and New Hope. Natural Foods Merchandiser., 1998: 14-16

Zochling, J., March, L., Lapsley, H., Cross, M., Tribe, K., and Brooks, P. 2004. Use of complementary medicines for osteoarthritis—a prospective study. Ann Rheum Dis., (2004), 63(5):549-54.

10. Green Leafy Vegetables

Leafy vegetables are rich sources of provitamin-A, vitamin C and minerals like calcium, magnesium, iron, *etc.*, They are the riches sources of roughes or fibres essential in human diet. The ICMR recommended dietary allowances of green leafy vegetables for adult women is 100g/day and men are 40g/day while for school children above 10 years it is 50g/day. Greens are rich in vitamins and minerals essential for healthy functioning of organs of body. Amaranth, beet leaf, spinach, fenugreek, *etc.*, are the major leaf vegetables grown in India. In addition, a number of under-utilized annual crops are also grown as leaf vegetables in specific regions. Tender stems and leaves of a number of perennial crops are rich sources of vitamins and minerals and are used for cooking.

10.1. Amaranth (*Amaranthus* sp)

Amaranth is the most common leaf vegetable grown in India. Amarnath is an annual herb with erect growth and scarce to profusely branching habit. Stem is succulent and green or purple or mixed shades of these two. There are two sections in Amaranthus, viz., *Amaranthus* and *Blitopsis*. Majority of leaf cultivars grown in India belong to *Amaranthus tricolor*. Among the several kinds of greens *Amaranthus* occupy a special place because it has a large number of species grown in one or the other part of country.

Some of the important improved varieties released are CO-2 and CO-5 from TNAU, Coimbatore, Pusa Kiran, Pusa Kirti and Pusa Lal Choulia from IARI, New Delhi, Arka Arunima and Arka Suguna from IIHR, Bangalore. Amaranth are popular all over India by different names. While in northern parts it is popularly known as choulia, in south it is popularly called as keerai or cheera or koora or soppa. The general composition of amaranth is given below:

There are several species of Amaranthus having different local names. Significant among them are:

Amaranthus tricolor (green and red type as well as Thandu Keerai or the succulent stems) (Syn. *Amaranthus gangeticu, A. mangostanus, A. polygonoides)*

Amarnathus dubius

Amaranthus spionosus (Spiny Amaranth)

Amaranthus tristis (Arai Keerai)

Amaranthus paniculatus (Rajagiri leaves) Grain Amaranth

Amaranthus polygonoides (Siru Keerai)

Amaranthus blitum (Siru keerai)

Amaranthus hypochondriachus (Dual purpose grain and leaf) Mola Keerai

Nutritional components of each of these species of Amaranth leaves

Sl. No.	Species	Energy Kcal/100g	Moisture (%)	Protein (%)	Fats (%)	Carbo-Hydrates (%)	Fibre (%)	Minerals (%)	Calcium (mg)	Phospho-rus (mg)	Iron (mg)
1.	Amaranthus Caudatus	26	90	3	1	2	1	3	200	40	-
2.	Amaranthus Gangeticus (leaves)	45	86	4	0	6	1	3	387	83	3
3.	Amaranthus Gangeticus (stem)	19	92	1	0	3	1	2	260	30	2
4.	Amaranthus polygonoides	33	90	3	0	5	-	2	251	55	27
5.	Amaranthus paniculatus	67	79	6	1	9	2	4	530	60	18
6.	Amaranthus spinosus	43	85	3	0	7	1	4	800	50	23
7.	Amaranthus Sp. (Chakravarti keerai)	57	81	4	1	8	2	4	321	71	18
8.	Amaranthus Sp. (Koya keerai)	37	88	3	0	5	2	1	292	51	2
9.	Amaranthus trisltis	44	87	3	0	7	-	2	364	62	38
10.	Amaranthus viridis	38	82	5	0	4	6	3	330	52	19

Phytochemicals

The leaves of amaranth contain besides the nutritional components, several other biomolecules of medicinal value like amarantin, isoamarantin and betanine. Besides iron they are also good source of copper and manganese. Amaranthus seeds have 30% higher protein than rice, wheat flour, oats, rye and other cereals. It has more calcium than milk. It contains amino acid 1-lysine, which is not found in plants. They contain other amino acids like arginine, cystine, histidine, leucine, isoleucine, methionine, phenylalanine, threonine, tryptophan and tyrosine. Bioassay-directed isolation of leaves and stems of *A. tricolor* yielded three galactosyl diacylglycerols (1-3) with potent cyclooxygenase and human tumor cell growth inhibitory activities (Jayaprakasham *et al.*, 2004).

Health Benefits of Amaranth Based on Traditional Knowledge and Folklore Practice

In traditional medicine, the roots of amaranth are used against colic, gonorrhoe and eczema. Amaranth has galactogogue properties. A mixture of boiled roots of amaranth and pulses is given to cow to yield more milk. The roots are demulcent, and decoction of roots is useful in treating strictures and piles and also in diarrhea of children. The leaves are useful in treating intestinal and urinary discharges. The roots and seeds are given in leucorrhea and impotence. The stems of amaranth are good for cough and bronchitis. It is a blood purifier and good tonic for dropsy or fluid retention. It is used as ascariside and is also useful in toothache and sore throat. Poultice of amaranth root is applied externally in ulcerated conditions of throat and mouth and as wash for ulcers. The whole plant is used as an antidote for snake poison and roots specifically for colic. Regular use of amaranth in our food items prevents the deficiency of vitamins A, B1, B2, and C, calcium, iron and potassium. It protects against several disorders such as defective vision, respiratory infections, recurrent colds retarded growth and functional sterility. In traditional and folklore medicine, amaranth is recommended with several other ingredient to correct common ailments, which are enlisted and elaborated below.

Respiratory Disorders: Amaranth is valuable in respiratory system disorders. Drinking fresh juice along with honey is remedy for chronic bronchitis, asthma, emphysema and tuberculosis.

Pregnancy, Childbirth and Lactation: Regular use of amaranth during pregnancy and lactation is highly beneficial. One cup of fresh leaf juice of amaranth mixed with honey and a pinch of cardamom powder should be taken during the entire period of pregnancy. It will help the normal growth of the baby, prevent the loss of calcium and iron from the body, relax the uterine ligaments and facilitate easy delivery without much pain. Its use after childbirth will shorten the laying in period, check the postnatal complications, and increase the flow of breast milk.

Retarded Growth: Amaranth is very useful in preventing retarded growth in children. A teaspoonful of the fresh juice mixed with few drops of honey should be given once every day to infants after a fort night of the birth. It will help the baby to grow healthy and strong. It will prevent constipation and ease the teething process as the baby grows. Growing children can safely be given this juice as a natural protein tonic. It contains all the essential amino acids such as arginine, nistidine, isoleucine, leucine, lysine, cystine, methionine, phenylalanine, threonine, tryptophan and valine.

Premature Ageing: The regular use of amaranth is useful in preventing premature aging. It prevents the disturbance of calcium and iron metabolism which usually occurs in old age. According to Dr. Van-Sylke, calcium molecules begin to get deposited in the bone tissues as one becomes old. This haphazard calcium distribution is influenced by the improper molecular movements of iron in the tissues. If this molecular disturbance of calcium and iron is prevented by regular supply of food. Calcium and iron as found in amaranth and the health is maintained by its regular use from the early age, the process of ageing can be prevented (Van Slykes, 1951).

Bleeding Tendencies: The use of amaranth is valuable in all bleeding tendencies. A cupful of fresh leaf juice mixed with a teaspoonful of lime juice should be taken every night in conditions like bleeding from the gums, nose, lungs, piles and excessive menstruation. It also acts as a natural tonic.

Leucorrhoea: Amaranth is beneficial in the treatment of leucorrhoea. The rind of the root of amaranth is rubbed in 250 ml. of water and strained, the strained juice is to be consumed twice daily (morning and evening). The roots have tendency to be eaten way by moths when stored for long. In case good root is not available, its leaves and branches may be used.

Venereal Diseases :- Amaranth is considered highly beneficial in the treatment of gonorrhea. About 25 gm's of the leaves of this vegetable should be given twice or thrice a day to the patient in case venereal diseases.

Therapeutic Properties of Amaranth

Devraj and Krishna (2011) evaluated the antiulcer activity of leaf extracts of *Amaranthus tricolor* Linn. (Amaranthaceae) in rats by inducing ulcers by five different means to see if extracts of amaranth have any cytoprotective property. The results revealed that it has very good antiulcer property in the experimental animal models which is consistent with the literature report in folk medicine. In another study, the ethanolic extract of *Amaranthus tricolor* L. (ATE) leaves when tested for its efficacy against chloroform-induced liver toxicity in rats, it clearly showed that oral administration of ATE for three weeks significantly reduced the elevated levels of serum GOT, GPT, GGT, ALP, bilirubin, cholesterol, LDL, VLDL, TG, and MDA suggesting its hepatoprotective property (Al-Dosari, 2010).

Jayaprakasam *et al*., (2004) tested three galactosyl diacylglycerols (1-3) compounds from amaranth for its antiproliferative activity using human AGS (gastric), CNS (central nervous system; SF-268), HCT-116 (colon), NCI-H460 (lung), and MCF-7 (breast) cancer cell lines, and found that compound 1 inhibited the growth of AGS, SF-268, HCT-116, NCI-H460, and MCF-7 tumor cell lines with IC50 values of 49.1, 71.8, 42.8, 62.5, and 39.2 mug/mL, respectively. For AGS, HCT-116, and MCF-7 tumor cell lines, the IC50 values of compounds 2 and 3 were 74.3, 71.3, and 58.7 microg/mL and 83.4, 73.1, and 85.4, respectively. This is the first report of the COX enzyme inhibitory activity for galactosyl glycerols and antiproliferative activities against human colon, breast, lung, stomach, and CNS tumor cell lines.

References

Al-Dosari, M. S. 2010. The effectiveness of ethanolic extract of *Amaranthus tricolor* L.: A natural hepatoprotective agent. Am J Chin Med. 2010;38(6):1051-64.

Devaraj, V.C., and Krishna, B.G. 2011.Gastric antisecretory and cytoprotective effects of leaf extracts of *Amaranthus tricolor* Linn. In rats. Zhong Xi Yi Jie He Xue Bao., (2011), 9(9):1031-8.

Jayaprakasam, B., Zhan,g Y., and Nair, M.G. 2004. Tumor cell proliferation and cyclooxygenase enzyme inhibitory compounds in *Amaranthus tricolor*. J Agric Food Chem., (2004), 52(23):6939-43.

Van Slykes. 1951. The Lancet, 258(6686): 723.

10.2. Spinach (Vilayathi Palak) (*Spinacia oleracea L*)

Spinach originated in central Asia, probably in Iran. Being a member of Chenopodiaceae family its similar to palak, sugar beet and Swiss Chard but is quite distinct. Spinach known as 'Vilayathi Palak' is typically a cool season leaf vegetable, and is widely cultivated in USA, Canada and Europe. In India its cultivation is limited to hilly regions or is available in plains during winters. Leaves are rich source of vitamin A and C and iron and calcium.

Nutritional Components of Spinach leaves and stalk

Spinach Leaves		Spinach Stalk	
Constituents	Quantity (per 100 g)	Constituents	Quantity (per 100 g)
Moisture	92 g	Moisture	93 g
Carbohydrates	3 g	Carbohydrates	4 g
Proteins	2 g	Proteins	1 g
Fats	0.1 g	Fats	0.00
Energy	26 Kcal	Energy	20 Kcal
Dietary Fibre	1 .2 g	Dietary Fibre	1.4 g
Iron	1.00 mg	Iron	2 mg
Phosphorus	21 mg	Phosphorus	20 mg
Calcium	73 mg	Calcium	90 mg

Goplan, *et al*., (1985)

Phytochemicals in Spinach and its Health Benefits

Spinach is considered to be a rich source of iron. It's loaded with vitamins and minerals, some of which are hard to find in other foods. Spinach is an excellent source of beta-carotene, a powerful disease-fighting antioxidant. It is also found to reduce the risk of developing cataract. It is also and excellent source of leutin. Lutein provides protection to the eyes from free radical damage caused by UV light rays from sun. It fights heart disease and cancer as well. Lutein is an important phyto-chemical found in spinach which helps prevent age related macular degeneration. The human body converts Beta-carotene into Vitamin A in the intestine and liver. Vitamin A is essential for eye health and vision, particularly prevention of night blindness, it assists in growth and formation of bone, maintains healthy hair, skin and mucus membrane, it also prevents and fights viral and respiratory diseases. Spinach also contains lipoic acid, which helps antioxidants vitamin C & E to regenerate. Because of its role in energy production, lipoxic acid is being investigated for regulating blood sugar levels. Spinach is the best natural source of Alpha Lipoic acid, a potent antioxidant and cancer fighter. It also boosts glutathione levels and regulates blood sugar levels. Glutathione is an antioxidant, and enzyme co-factor. It is required by the human body to maintain normal immune system function. Low glutathione levels are known to be associated with conditions such as Parkinson's disease, Alzheimer's disease, multiple selerosis and male infertility. The potassium is available in abundance in spinach which helps in promotion of heart health. Spinach offers twice as much fibre as other greens.

Spinach is rich in betaine, a derivative of choline, which lowers homocysteine levels and thus reduces the risk of cardiovascular disease. The dark green colour of spinach is due to chlorophyll-a, compound having antioxidant property. It helps to cleanse and decolorize the human gastrointestinal system. It acts as a mild antibiotic and recent findings have shown that it has powerful anti-cancer properties. Chlorophylin, which is chlorophyll derivative, prevents RNA & DNA damage. Spinach is one of the better sources of dietary vitamin K whose main function is to control blood clotting. It has now been found that Vitamin K is necessary for bone health because it glues calcium into the bone matrix.

Maeda *et al.*, (2008) succeeded in purifying a major glycolipid fraction from a green spinach. This fraction consists mainly of three glycolipids: monogalactosyl diacylglycerol (MGDG), digalactosyl diacylglycerol (DGDG), and sulfoquinovosyl diacylglycerol (SQDG). They suggest that the orally administered glycolipid fraction from spinach could suppress colon tumor growth in mice by inhibiting the activities of neovascularization and cancer cellular proliferation in tumor tissue.

Daily consumption of fruits and vegetables is frequently recommended to prevent several diseases. This health-promoting effect is considered to be in part due to the antioxidant content of fruits and vegetables and their ability to decrease oxidative stress. Schirrmacher *et al.*, (2010) quantified antioxidant levels in plasma, kinetics of lipid peroxidation, MDA concentration, and total antioxidative capacity of plasma after feeding human volunteers with spinach preparation for 10 days (5 mg lutein/d). They observed a significant increase in lutein content and a moderate increase (n.s.) in beta-carotene content in human blood plasma after consumption of spinach or perilla. The markers of lipid peroxidation tended to decrease, but no influence on antioxidative capacity of plasma could be detected. The high lutein content of perilla caused a more pronounced increase of lutein compared to spinach. Both vegetables seem to be able to influence lipid peroxidation in a beneficial manner.

Several natural benefits and therapeutic properties of spinach has been mentioned in folklore medicine which are enumerated below.

- The leaves of spinach are demulcent or soothing agents, refrigerant or collants, diuretic and milk laxative
- **Constipation**: Spinach juice cleans the digestive tract by removing the accumulated digestive waste. It nourishes the intestines and tones up their movements. It is, therefore, an excellent food remedy for constipation.
- **Anemia**: Spinach is a valuable source of high-grade iron that is more ioavailable. Iron helps in the formation of hemoglobin in red blood cells.
- **Acidosis**: Spinach is also a rich source of calcium and other alkaline elements which are essential for keeping the tissues clean and for preserving the alkalinity of the blood. It, therefore, helps prevent chronic diseases which thrive on the formation of too much acid in the system.
- **Night Blindness**: The spinach is particularly rich in vitamin A. It contains more vitamin A than most other green vegetables. Lack of vitamin A may lead to night

blindness. Spinach is thus an effective food remedy for the prevention and treatment or night blindness.

- **Tooth Disorders**: The spinach juice is effective in strengthening the gums and preventing and curing dental caries. Chewing raw spinach leaves cures pyorrhoea. A mixture of carrot juice and spinach juice, taken early in the morning, can cure bleeding and ulcerated gums.

- **Pregnancy and Lactation**: Spinach is the richest source of folic acid, that is essential during pregnancy and lactation. Megaloblastic anemia in pregnant women occurs because of deficiency of folic acid. The deficiency of folic acid occurs as this substance is required for the developing foetus. Regular use of spinach during pregnancy will help prevent the deficiency of folic acid. It will also prevent miscarriage, accidental haemorrhage, improper absorption of food nutrients by small intestine associated with lassitude i.e. tiredness, shortness of breath, loss of weight and diarrhea. Spinach is also good source of nutrition for nursing or lactating mothers and improves the quality of their milk.

- **Urinary Disorders**: Fresh spinach juice taken with tender coconut water once or twice a day acts as a very effective but safe diuretic due to the combined action of both nitrates and potassium. It can be safely given in cystitis, nephritis and scanty urination due to dehydration.

- **Respiratory Disorders**: Infusion of fresh leaves of spinach prepared with two teaspoonful of fenugreek seeds mixed with a pinch of ammonium chloride and honey is an effective expectorant tonic during the treatment of bronchitis, tuberculosis, asthma and dry cough due to congestion in the throat. It soothens the bronchioles, liquefies the sputum and forms healthy tissues in the lungs and increase resistance against respiratory infections. It should be taken in doses of 30 ml. three times daily.

References

Bose, T.K. and Som, M.G. 1986. Vegetable crops in India. Naya Prokash, 206, Bidhan Sarani, Calcutta.

Gopalan, C., Sastri, B.V.R., and Balasubramanian, S.C. 1985. Nutritive Value of Indian Foods. National Institue of Nutrition, ICMR, Hyderabad, India.

Maeda, N., Kokai, Y., Ohtani, S., Sahara, H., Kumamoto-Yonezawa, Y., Kuriyama, I., Hada, T., Sato, N., Yoshida, H., Mizushina, Y. 2008. Anti-tumor effect of orally administered spinach glycolipid fraction on implanted cancer cells, colon-26, in mice., (2008), 43(8):741-748.

Parasramka, M.A., Dashwood, W.M., Wang, R., Abdelli, A., Bailey, G.S., Williams, D.E., Ho, E., and Dashwood, R.H. 2012. MicroRNA profiling of carcinogen-induced rat colon tumors and the influence of dietary spinach. Mol Nutr Food Res., (2012), 56(8):1259-69.

Schirrmacher, G., Skurk, T., Hauner, H., and Grassmann, J. 2010. Effect of *Spinacia oleraceae* L. and *Perilla frutescens* L. on antioxidants and lipid peroxidation in an intervention study in healthy individuals. Plant Foods Hum Nutr., (2010), 65(1):71-6.

10.3. Fenugreek (*Trigonella foenum-graecum*)

Fenugreek (*Trigonella foenum-graecum*) is a plant in the family Fabaceae. Fenugreek is used both as a herb (the leaves) and as a spice (the seed). The leaves, known as *methi*, are also eaten as vegetable. The plant is cultivated worldwide as a semi-arid crop and is a common ingredient in dishes from the Indian Subcontinent.

India, Pakistan, Nepal, Bangladesh, Argentina, Egypt, France, Spain, Turkey, Morocco and China are the major fenugreek producing countries. India is the largest producer of fenugreek in the world. In India it is produced widely in Rajasthan, Gujarat, Uttaranchal, Uttar Pradesh, Madhya Pradesh, Maharashtra, Haryana and Punjab. Rajasthan contributes 80% of the total production of fenugreek in India. Qasuri Methi, more popular for its appetizing fragrance, comes from Qasur in Pakistan.

Nutritional Components of Fenugreek Leaves

Constituent	Quantity (per 100g)	Constituent	Quantity (per 100g)
Moisture	86.1 g	Phosphorus	51 mg
Carbohydrates	6.0 g	Iron	16.5 mg
Proteins	4.4 g	Carotene	2340 µg
Fat	0.9 g	Thiamine	0.04 mg
Fibre	1.1 g	Riboflavin	0.31 mg
Minerals	1.5 g	Niacin	0.8 mg
Calcium	395 mg	Vitamin C	52 mg

Nutritional Components of Fenugreek Seeds

Constituent	Quantity (per 100g)	Constituent	Quantity (per 100g)
Moisture	13.700 gm	Iron	6.500 mg
Protein	26.200 gm	Carotene	96.000 µg
Fat	5.800 gm	Thiamine	0.340 mg
Minerals	3.000 gm	Riboflavin	0.290 mg
Fibre	7.200 gm	Niacin	1.100 mg
Carbohydrates	44.100 gm	Folic Acid (Total	84.000 µg
Energy	333.000 K cal	Choline	1161.000 mg
Calcium	160.000 mg	Copper	0.710 mg
Magnesium	124.000 mg	Manganese	1.030 mg
Sodium	19.000 mg	Zinc	3.080 mg
Potassium	530.000 mg	Chromium	0.064 mg
Phosphorus	370.000 mg	Phytin Phosphorus	151.000 mg

Gopalan *et al.*, (1985)

Phytochemicals in Fenugreek and Their Health Benefits

Fenugreek seeds are a rich source of the polysaccharide galactomannan. They are also a source of saponins such as diosgenin, yamogenin, gitogenin, tigogenin, and neotigogens. Other bioactive constituents of fenugreek include mucilage, volatile oils, and alkaloids such as choline and trigonelline. 4-hydroxyleucine, a novel amino acid found in fenugreek seeds which increases glucose stimulated insulin secretion.

Fenugreek seeds are used as a medicinal in Traditional Chinese Medicine under the name Hu Lu Ba (Traditional Chinese), where they are considered to warm and tone up kidneys, disperse cold and alleviate pain. They are used raw or toasted. In India about 2-3g of raw fenugreek seeds are swallowed raw early in the morning with warm water, before brushing the teeth and before drinking tea or coffee, where they are supposed to have a therapeutic and healing effect on joint pains, without any side effects. Drinking 1 cup of fenugreek tea per day, made from the leaves, is said to relieve the discomfort of arthritis. Fenugreek seeds are thought to be a galactagogue that is often used to increase milk supply in lactating women. Its seeds have been used in many traditional medicines as laxative, digestive, and as a remedy for cough and bronchitis.

If used regularly, fenugreeks may help control cholesterol, triglyceride as well as high blood sugar (glycemic) levels in diabetics. Fenugreek seeds is lactogogue and if added to cereals and wheat flour (bread) or made into gruel and given to the nursing mothers, it increases lactation.

• Fenugreek seeds are rich source of minerals, vitamins, and phytonutrients.

• The seeds are very good source of soluble dietary fiber. Soaking the seeds in water makes their outer coat soft and mucilaginous. 100 g of seeds provide 24.6 g or over 65% of dietary fiber.

• Non-starch polysaccharides (NSP) which constitute major fiber content in the fenugreeks include *saponins, hemicelluloses, mucilage, tannin,* and *pectin*. These compounds help lower blood LDL-cholesterol levels by inhibiting bile salts re-absorption in the colon. They also bind to toxins in the food and helps to protect the colon mucus membrane from cancers.

• NSPs (non-starch polysaccharides) increase the bulk of the food and augments bowel movements. Altogether, NSPs assist in smooth digestion and help relieve constipation ailments.

• It has been established that amino-acid, 4-hydroxy isoleucine present in the fenugreek seeds facilitate action on insulin secretion. In addition, fiber in the seeds help lower rate of glucose absorption in the intestines and thus controls blood sugar levels. The seeds are therefore recommended in diabetic diet.

• The seeds contain many phytochemical compounds such as *choline, trigonelline diosgenin, yamogenin, gitogenin, tigogenin* and *neotigogens*. Together, these compounds account for the medicinal properties of fenugreeks.

• The fenugreek seeds are an excellent source of minerals like copper, potassium, calcium, iron, selenium, zinc, manganese, and magnesium. Potassium is an important component of cell and body fluids that helps control heart rate and blood pressure by countering action on sodium. Iron is essential for red blood cell production and as a co-factor for cytochrome-oxidase enzymes.

• It is also rich in many vital vitamins including thiamin, pyridoxine (vitamin B-6), folic acid, riboflavin, niacin, vitamin-A and vitamin-C that are essential nutrients for optimum health.

Therapeutic Properties of Fenugreek

Diabetes mellitus (DM) is a metabolic disorder caused by insufficient or inefficient insulin secretary response and it is characterized by increased blood glucose levels (hyperglycemia). Glucose is the main energy source for the body, and in the case of DM, management of glucose becomes irregular. There are three key defects in the onset of hyperglycemia in DM, namely increased hepatic glucose production, diminished insulin secretion, and impaired insulin action. Conventional drugs treat diabetes by improving insulin sensitivity, increasing insulin production and/ or decreasing the amount of glucose in blood. Diabetes mellitus is a heterogeneous metabolic disorder characterized by hyperglycaemia resulting in defective insulin secretion, resistance to insulin action or both. The use of biguanides, sulphonylurea and other drugs are valuable in the treatment of diabetes mellitus. Their use, however, is restricted by their limited action, pharmacokinetic properties, secondary failure rates and side effects. *Trigonella foenum-graecum*, commonly known as fenugreek, is a plant that has been extensively used as a source of antidiabetic compounds from its seeds and leaf extracts. Preliminary human trials and animal experiments suggest possible hypoglycaemic and antihyperlipedemic properties of fenugreek seed powder taken orally.

Studies by Xue *et al.*, (2011) concluded that Trigonella Extract (TE) confers protection against functional and morphologic injuries in the kidneys of diabetic rats by increasing activities of antioxidants and inhibiting accumulation of oxidized DNA in the kidney, suggesting a potential drug for the prevention and therapy of DM. Middha *et al.*, (2011) observed that the potency of fenugreek in restoring several parameters to normal values is comparable to glibenclamide (an insulin regulating drug).

Fenugreek Seed Powder significantly alleviated most signs of the metabolic syndrome resulting from experimentally induced type 1 diabetes and obesity by 40-76 and 56-78 %, respectively, including hyperglycaemia, hyperlipidaemia, elevation in atherogenic indices, impairment of liver functions, severe changes in body weight and oxidative stress. Besides, FSP (especially the high dose) completely modulated the immunosuppressive activity of CP (an immuno-suppresor) including leucopenia (resulting from neutropenia and lymphopenia), decrease in weights and cellularity of lymphoid organs, serum γ-globulin level, delayed type of hypersensitivity response and delay in the skin-burn healing process (Ramadan *et al.*, 2011). They concluded that fenugreek seeds may be useful not only as a dietary adjunct for the control of the metabolic syndrome in diabetic/obese patients, but also as an immune-stimulant in immune-compromised patients such as those under chemotherapeutic interventions.

GII is a water-soluble compound purified from fenugreek seeds that exhibit anti-diabetic property. The studies of Moorthy *et al.*, (2010) revealed that GII purified from fenugreek (T. *foenum-graecum*) seeds decreased lipid content of liver and stimulated the enzymes of glycolysis (except glucokinase) and inhibited enzymes of gluconeogenesis in the liver of the diabetic especially moderately diabetic rabbits. Findings of Uemura *et al.*, (2011) suggest that fenugreek ameliorates dyslipidemia by decreasing the hepatic lipid content in diabetic mice and that its effect is mediated by diosgenin. Fenugreek, which contains diosgenin, may be useful for the management of diabetes-related hepatic dyslipidemias.

Hamden et al., (2010) observed that fenugreek steroids (designated as F-steroids) administration to surviving diabetic rats significantly decreased the sperm shape abnormality and improved the sperm count. Above all, the potential protective action of reproductive systems was approved by the histological study of testis and epididymis, suggesting its role in reproductive health.

Fenugreek oil was also found to have several health benefits. The study of Hamden et al., (2010) revealed the efficacy of fenugreek oil in the amelioration of diabetes, hematological status, and renal toxicity which may be attributed to its immunomodulatory activity and insulin stimulation action along with its antioxidant potential.

Kassaian et al., (2009), in a clinical trial, placed 24 type 2 diabetic patients on 10 grams/day powdered fenugreek seeds mixed with yoghurt or soaked in hot water for 8 weeks. This study showed that fenugreek seeds can be used as an adjuvant in the control of type 2 diabetes mellitus in the form of soaked in hot water.

The blood glucose level lowering effect of fenugreek was found to be almost comparable to the effect of insuli in a study conducted by Baquer et al., (2011). Combination with trace metal showed that vanadium had additive effects and manganese had additive effects with insulin on in-vitro system in control and diabetic animals of young and old ages using adipose tissue. The Trigonella and vanadium effects were studied in a number of tissues including liver, kidney, brain peripheral nerve, heart, red blood cells and skeletal muscle. Addition of Trigonella to vanadium significantly removed the toxicity of vanadium when used to reduce blood glucose levels.

A study by Yadav et al., (2010) showed that water extract of T. graecum seeds have higher hypoglycemic and antihyperglycemic potential and may of use as complementary medicine to treat the diabetic population by significantly reducing dose of standard drugs.

Sauvaire et al., (1998) reported characterization of a new insulinotropic compound, 4-hydroxyisoleucine extracted and purified from fenugreek. They also showed that 1) the pattern of insulin secretion induced by 4-hydroxyisoleucine was biphasic, 2) that this effect occurred in the absence of any change in pancreatic alpha- and delta-cell activity, and 3) that the more glucose concentration was increased, the more insulin response was amplified. Moreover, 4-hydroxyisoleucine did not interact with other agonists of insulin secretion (leucine, arginine, tolbutamide, glyceraldehyde).

Jette et al., (2009) reported that the major isomer of 4-hydroxyisoleucine, an atypical branched-chain amino acid derived from fenugreek, is responsible for the effects of this plant on glucose and lipid metabolism. 4-Hydroxyisoleucine was demonstrated to stimulate glucose-dependent insulin secretion by a direct effect on pancreatic islets. In addition to stimulating insulin secretion, 4-hydroxyisoleucine reduced insulin resistance in muscle and/or liver by activating insulin receptor substrate-associated phosphoinositide 3 (PI3) kinase activity. 4-Hydroxyisoleucine also reduced body weight in diet-induced obese mice. The decrease in body weight was associated with a marked decrease in both plasma insulin and glucose levels, both of which are elevated in this animal model. Finally, 4-hydroxyisoleucine decreased elevated plasma triglyceride and total cholesterol levels in a hamster model of diabetes. Based on the beneficial metabolic properties that have been demonstrated, 4-hydroxyisoleucine, a simple, plant-derived amino acid, may represent an attractive new candidate for the treatment of type 2 diabetes, obesity and dyslipidemia, all key components of metabolic syndrome.

Hannan et al., (2007) indicated that daily oral administration of Soluble Dietary Fibre (SDF) fraction of fenugreek seeds to type 2 diabetic rats for 28 days decreased serum glucose, increased liver glycogen content and enhanced total antioxidant status. Serum insulin and insulin secretion were not affected by the SDF fraction. Glucose transport in 3T3-L1 adipocytes and insulin action were increased by *T. foenum-graecum*. Their findings indicated that the SDF fraction of *T. foenum-graecum* seeds exerts antidiabetic effects mediated through inhibition of carbohydrate digestion and absorption, and enhancement of peripheral insulin action.

References

Baquer, N.Z., Kumar, P., Taha, A., Kale, R.K., Cowsik, S.M., and McLean, P. 2011. Metabolic and molecular action of *Trigonella foenum-graecum* (fenugreek) and trace metals in experimental diabetic tissues. J Biosci., (2011), 36(2):383-96.

Bose, T.K. and Som, M.G. 1986. Vegetable crops in India. Naya Prokash, 206, Bidhan Sarani, Calcutta.

Gopalan, C., Sastri, B.V.R., and Balasubramanian, S.C. 1985. Nutritive Value of Indian Foods. National Institue of Nutrition, ICMR, Hyderabad, India.

Hamden, K., Jaouadi, B., Carreau, S., Aouidet, A., El-Fazaa, S., Gharbi, N., and Elfeki A. 2010. Potential protective effect on key steroidogenesis and metabolic enzymes and sperm abnormalities by fenugreek steroids in testis and epididymis of surviving diabetic rats. Arch Physiol Biochem., (2010), 116(3):146-55.

Hamden, K., Masmoudi, H., Carreau, S., and Elfeki A. 2010. Immunomodulatory, beta-cell, and neuroprotective actions of fenugreek oil from alloxan-induced diabetes. Immunotoxicol., (2010), 32(3):437-45.

Hannan, J.M., Ali, L., Rokeya, B., Khaleque, J., Akhter, M., Flatt, P.R., and Abdel-Wahab, Y.H. 2007. Soluble dietary fibre fraction of *Trigonella foenum-graecum* (fenugreek) seed improves glucose homeostasis in animal models of type 1 and type 2 diabetes by delaying carbohydrate digestion and absorption and enhancing insulin action. Br J Nutr., (2007), 97(3):514-21.

Jetté, L., Harvey, L., Eugeni, K., and Levens, N. 2009. 4-Hydroxyisoleucine: a plant-derived treatment for metabolic syndrome, (2009), 10(4):353-8.

Kassaian, N., Azadbakht, L., Forghani, B., and Amini, M. 2009. Effect of fenugreek seeds on blood glucose and lipid profiles in type 2 diabetic patients. Int J Vitam Nutr. Res., (2009), 79(1):34-9.

Middha, S.K., Bhattacharjee, B., Saini, D., Baliga, M.S., Nagaveni, M.B., Usha, T. 2011. Protective role of *Trigonella foenum graceum* extract against oxidative stress in hyperglycemic rats. Eur Rev Med Pharmacol Sci., (2011), 15(4):427-35.

Moorthy, R., Prabhu, K.M., and Murthy, P.S. 2010. Mechanism of anti-diabetic action, efficacy and safety profile of GII purified from fenugreek (*Trigonella foenum-graceum* Linn.) seeds in diabetic animals. Indian J Exp Biol., (2010), 48(11):1119-22.

Ramadan, G., El-Beih, N.M., Aand bd El-Kareem, H.F. 2011. Anti-metabolic syndrome and immunostimulant activities of Egyptian fenugreek seeds in diabetic/obese and immunosuppressive rat models. Br J Nutr., (2011), 105(7):995-1004.

Sauvaire, Y., Petit, P., Broca, C., Manteghetti, M., Baissac, Y., Fernandez-Alvarez, J., Gross, R., Roye, M., Leconte, A., Gomis, R., and Ribes, G. 1998. 4-Hydroxyisoleucine: a novel amino acid potentiator of insulin secretion. Diabetes, (1998), 47(2):206-10.

Uemura, T., Goto, T., Kang, M.S., Mizoguchi, N., Hirai, S., Lee, J.Y., Nakano, Y., Shono, J., Hoshino, S., Taketani, K., Tsuge, N., Narukami, T., Makishima, M., Takahashi, N., and Kawada, T. 2011. Diosgenin, the main aglycon of fenugreek, inhibits LXRα activity in HepG2 cells and decreases plasma and hepatic triglycerides in obese diabetic mice. J Nutr., (2011), 141(1):17-23.

Xue, W., Lei, J., Li, X., and Zhang, R. 2011. *Trigonella foenum graecum* seed extract protects kidney function and morphology in diabetic rats via its antioxidant activity. Nutr Res., (2011), 31(7):555-62.

Yadav, M., Lavania, A., Tamar, R., Prasad, G.B., Jain, S., and Yadav, H. 2010. Complementary and comparative study on hypoglycemic and antihyperglycemic activity of various extracts of Eugenia jambolana seed, Momordica charantia fruits, Gymnema Sylvester and Trigonella foenum graecum seeds in rats. Appl Biochem Biotechnol., (2010), 160(8): 2388-2400. Dio: 10.1007/s12010-009-8799-1

10.4. Moringa Leaves (*Moringa oleifera* L.)

Moringa oleifera Lam (syn. M. ptreygosperma Gaertn.) belongs to a monogeneric family Moringaceae (Nadkarni, 1976; Ramachandran *et al.*, 1980). There are 14 species in Moringa genus, of which *Moringa oleifera* and *Moringa stenopetala* are commercially important. Though native to foothills of Himalayas, it is now cultivated widely in Africa, Central and South America, India, Srilanka, Malaysia and the Phillippines (Morton, 1991; Mughal *et al.*, 1999). In some parts of the world *M. oleifera* is referred to as the 'drumstick tree' or the 'horse radish tree'. It is commonly known in India as sahjan, muga, munga, muringakkai, muringakkaya, munagakaya, nuggekai, sajane dauta, saragavo, shevaga, drumstick, horse radish tree *etc.*, The leaves, fruit, flowers and immature pods of this tree are used as a highly nutritive vegetable in many countries, particularly in India, Pakistan, Philippines, Hawaii and many parts of Africa (Anwar *et al.*, 2007). Moringa leaves have been reported to be a rich source of β -carotene, protein, vitamin C, calcium and potassium and act as a good source of natural antioxidants. It contains various types of antioxidant compounds such as ascorbic acid, flavonoids, phenolics and carotenoids (Dillard and German, 2000; Siddhuraju and Becker, 2003). Because of its lactogogue property and high content of iron, it is known as 'mother's best friend' in the Philippines (Siddhuraju and Becker, 2003).

Cultivars: KM 1, Dhanraj, KDM 1, PKM 1 and PKM 2 are the popular moringa varieties/hybrids available in India.

Nutritional Components of Moringa Leaves

Constituent	Quanty per 100g	Constituent	Quantyity per 100g
Moisture	81.0 g	Iron	7 mg
Carbohydrates	14.3 g	Copper	110 µg
Proteins	6.70 g	Iodine	5.1 µg
Fats	1.7 g	Vitamin A	11,300 IU
Minerals	2.30 g	Vitamin B6	120 µg
Dietary Fibres	0.9 g	Nicotinic acid	0.80 mg
Calcium	440 mg	Vitamin C	220 mg
Phosphorus	70 mg	Tocopherol	7.4 mg

Gopalan *et al.*, (1985)

Moringa leaves contain seven times the Vitramin C in oranges, four times the calcium in milk, four times the Vitamin A in carrots and two times the protein in milk and three times the potassium in bananas.

Phytochemicals

Moringa is rich in glucosinolates and isothiocyanates (Bennet *et al.*, 2003). Its bark has two alkaloids moringine, moringinine, besides other compounds like vanillin, β-sitosterol, β-sitostenone, 4-hydroxymellin and octacosanoic acid (Anwar *et al.*, 2007). The sitosterol is an oestrogenic substance. The flowers are reported to contain nine amino acids, sucrose, D-glucose, wax, quercetin and kaemferol. Leaves of moringa is said to be the richest source of calcium, potassium and iron. This leaf also possesses phytochemicals like benzoyl isothioxcyanate and benzoyl glucosinolate that have anti-cancerous and anti-microbial properties.

Thiocarbamate and isothiocyanate glycosides have been isolated from the acetate phase of the ethanol extract of Moringa pods (Faizi et al., 1998), which are reported to have antihypertensive properties. The high concentrations of ascorbic acid, oestrogenic substances and β -sitosterol [16], iron, calcium, phosphorus, copper, vitamins A, B and C, α -tocopherol, riboflavin, nicotinic acid, folic acid, pyridoxine, β -carotene, protein, and in particular essential amino acids such as methionine, cystine, tryptophan and lysine are present in Moringa leaves and pods and make it a virtually ideal dietary supplement (Makkar and Becker, 1996).The cytokinins have also been reported to be present in the fruit (Nagar et al., 1982).

Health Benefits of Moringa (leaves & pods) and the Research Findings Supporting its Therapeutic Properties

Moringa confers several health benefits due to the presence of several phytochemicals that have antibiotic, antitrypanosomal, hypotensive, antispasmodic anti ulcer, anti inflammatory, hypocholes teromic and hypoglycemic activities. Moringa is reported to be a strong antioxidant effective against prostate and skin cancers, anti-tumor and anti-aging substance. It also modulates anemia, high blood pressure, diabetes, high serum or blood cholesterol, thyroid, liver and kidney problems. It has strong anti-inflammatory properties ameliorating rheumatism, joint pain, arthritis, edema and Lupus. It is effective against digestive disorders including colitis, diarrhea, flatulence, ulcer or gastritis. As a detoxifying agent, it is effective against snake and scorpion bites. It is effective against nervous disorders including headaches, migraines, hysteria and epilepsy.

Several in-vitro and in-vivo pharmacological studies conducted by researchers have validated the benefits of moringa and have also postulated probable mechanisms responsible for bringing out the beneficial effects. The crude extract of Moringa leaves has a significant cholesterol lowering action in the serum of high fat diet fed rats which might be attributed to the presence of a bioactive phytoconstituent, i.e. β -sitosterol (Ghasi et al., 2000). Moringa fruit has been found to lower the serum cholesterol, phospholipids, triglycerides, low density lipoprotein (LDL), very low density lipoprotein (VLDL) cholesterol to phospholipid ratio, atherogenic index lipid and reduced the lipid profile of liver, heart and aorta in hypercholesteremic rabbits and increased the excretion of fecal cholesterol (Mehta et al., 2003).

Ethanol extract of moringa leaves have been found to exhibit antispasmodic effects possibly through calcium channel blockade (Gilani et al., 1992; 1994; Dangi et al., 2002). The methanol fraction of M. oleifera leaf extract showed antiulcerogenic and hepatoprotective effects in rats (Pal et al., 1995). Aqueous leaf extracts also showed antiulcer effect (Pal et al., 1995) indicating that the antiulcer component is widely distributed in this plant. Moringa roots have also been reported to possess hepatoprotective activity (Ruckmani et al., 1998). Moringa contains an active antibiotic principle, pterygospermin, which has powerful antibacterial and fungicidal effects (Ruckmani et al., 1998). The juice from the stem bark showed antibacterial effect against Staphylococcus aureus (Mehta et al., 2003). The fresh leaf juice was found to inhibit the growth of microorganisms (Pseudomonas aeruginosa and Staphylococcus aureus), which are human pathogenic bacteria (Caceres et al., 1991). The seed extracts have also been found to be effective on hepatic carcinogen metabolizing enzymes, antioxidant parameters and skin papillomagenesis in mice (Bharali et al., 2003). It has been found that niaziminin, a thiocarbamate from the leaves of M. oleifera, exhibits inhibition of tumor-promoter-induced Epstein–Barr virus activation.

Because of such several health promoting properties of moringa, there is ever increasing demand for moringa leaf powder and fruit powders in export market.

References

Anwar, F., Latin, S., Ashraf, M., and Gilani, A.H. 2007. Moringa oleifera: A food plant with multiple medicinal uses. Phytother.Res., (2007), 21:17-25.

Bennett RN, Mellon FA, Foidl N et al., 2003 Profiling glucosinolates and phenolics in vegetative and reproductive tissues of the multi-purpose trees Moringa oleifera L. (Horseradish tree) and Moringa stenopetala L. J Agric Food Chem 51: 3546–3553.

Bharali, R., Tabassum, J., Azad, M.R.H. 2003. Chemomodulatory effect of Moringa oleifera, Lam, on hepatic carcinogen metabolizing enzymes, antioxidant parameters and skin papillomagenesis in mice. Asia Pacific J Cancer Prev., 4: 131–139.

Caceres, A., Cabrera, O., Morales, O., Mollinedo, P., and Mendia, P. 1991. Pharmacological properties of Moringa oleifera. 1: Preliminary screening for antimicrobial activity. J Ethnopharmacol 33: 213–216.

Dangi, S.Y., Jolly, C.I., and Narayana, S. 2002. Antihypertensive activity of the total alkaloids from the leaves of Moringa oleifera. Pharm Biol., 40: 144–148.

Dillard, C.J., and German, J.B. 2000. Phytochemicals: nutraceuticals and human health: A review. J Sci Food Agric., 80: 1744– 1756.

Faizi, S., Siddiqui, B.S., Saleem, R., Aftab, K., Shaheen, F., and Gilani, A.H. 1998. Hypotensive constituents from the pods of Moringa oleifera. Planta Med., 64: 225–228.

Ghasi, S., Nwobodo, E., and Ofili, J.O. 2000. Hypocholesterolemic effects of crude extract of leaf of Moringa oleifera Lam in high-fat diet fed Wistar rats. J Ethnopharmacol., 69: 21– 25.

Gilani, A. H, Aftab, K., and Shaheen, F. 1992. Antispasmodic activity of active principle from Moringa oleifera. In Natural Drugs and the Digestive Tract, Capasso F, Mascolo N (eds). EMSI: Rome, 60–63.

Gilani, A.H., Aftab, K., and Suria, A. 1994. Pharmacological studies on hypotensive and spasmodic activities of pure compounds from Moringa oliefera. Phytother. Res., 8:87-91.

Makkar, H.P.S., and Becker, K. 1996. Nutritional value and antinutritional components of whole and ethanol extracted Moringa oleifera leaves. Anim Feed Sci Technol., 63:211-228.

Mehta, L.K., Balaraman, R., Amin, A.H., Bafna, P.A., and Gulati, O.D. 2003. Effect of fruits of Moringa oleifera on the lipid profile of normal and hypercholesterolaemic rabbits. J Ethnopharmacol., 86: 191–195.

Morton, J.F. 1991. The horseradish tree, Moringa pterigosperma (Moringaceae). A boon to arid lands. Econ Bot., 45: 318–333.

Mughal, M.H., Ali, G., Srivastava, P.S., and Iqbal, M. 1999. Improvement of drumstick (*Moringa pterygosperma* Gaertn.) – a unique source of food and medicine through tissue culture. Hamdard Med 42: 37–42.

Nagar, P.K., Iyer, R.I, and Sircar, P.K. 1982. Cytokinins in developing fruits of *Moringa pterigosperma* Gaertn. Physiol Plant., 55: 45–50.

Pal, S.K., Mukherjee, P.K., and Saha, B.P. 1995. Studies on the antiulcer activity of *Moringa oleifera* leaf extract on gastric ulcer models in rats. Phytother.Res., 9:463-465.

Ruckmani, K., Kavimani, S., Anandan, R., and Jaykar, B. 1998. Effect of *Moringa oleifera* Lam on paracetamol-induced hepatoxicity. Indian J Pharm Sci., 60:33-35.

Siddhuraju, P., and Becker, K. 2003. Antioxidant properties of various solvent extracts of total phenolic constituents from three different agro-climatic origins of drumstick tree (*Moringa oleifera* Lam.). J Agric Food Chem., 15: 2144–2155.

Somali, M.A., Bajnedi, M.A., and Al-Faimani, S.S. 1984. Chemical composition and characteristics of *Moringa peregrina* seeds and seed oil. J Am Oil Chem., Soc 61: 85–86.

10.5. Basella (*Basella rubra* L.)

Basella is an important green leafy vegetable species of Chenopodiaceae family and is known for its medicinal properties. Variously called Malabar or Ceylon Spinach, Basale or Pasalai Keerai, Pui/Poi Saag, *etc.*, in the many Indian languages, Basella is a wonderful alternate for the regular spinach. Basella commonly known as poi, Malabar Nightshade or Indian spinach is a popular summer leafy vegetable grown in almost all parts of India. The other local names of this leafy vegetable is Maya-ki-baji (Gujarathi), Rukhtopori (Bengali), Bachali (Telugu), Pasalai keerai or Vaslakire (Tamil) and Pachala (Malayalam). It is used in other Asian cuisines as well. It grows readily during the hot summer months unlike the sensitive spinach which bolts into seeds at the barest whisper of warmth. It is a lovely plant that grows readily from cuttings or seeds. Basella is somewhat succulent and slightly mucilaginous. Once cooked, there does not seem to be any evidence of the stickiness. Basella is delicious in dal, molakootal, and soup; it workes very well in the rice salad too.

The fresh tender leaves and stems are consumed as leafy vegetables after cooking. The colouring matter present in the red cultivars is reported to have been used in China as a dye. On account of the presence of mucilaginous substances in the leaves and stems, it is used as poultice. The juice of leaves is prescribed in cases of constipation, particularly for children and pregnant women.

Nutritional components of Basella

Constituent	Quantity per 100 g
Moisture	93 g
Protein	1.2 g
Calcium	0.15 g
Iron	1.4 mg
Vitamin A	3250 IU
Vitamin C	80 mg
Phosphorus	83.5 mg
Magnesium	103 mg

Phytochemicals

Malabar Spinach is very nutritious and helpful in curing malnutrition. The leaves contain Vitamin A, Vitamin E, Vitamin K, flavonoids, saponins, β-Carotene, water soluble polysaccharides, bioflavonoids, essential amino acids (arginine, leucine, isoleucine, lysine, threonine and tryptophan) and minerals (rich in calcium and iron compounds and contains a low percentage of soluble oxalates). This herb works as medicine due to the presence of numerous biologically active compounds such as carbohydrates, proteins, enzymes, fats and oils, vitamins, alkaloids, quinines, terpenoids, flavonoids, carotenoids, sterols, simple phenolic glycosides, tannins, saponins, polyphenols.

Health benefits of *Basealla alba (rubra)*

Basella alba has exhibited several biological activities like androgenic, antidiabetic, anti-inflammatory, antimicrobial, antioxidant, anti-ulcer, antiviral, CNS depressant, and hepatoprotective. The plant was also used in traditional medicine for cure of digestive disorders, skin diseases, bleeding piles, pimples, urticaria, irritation, anemia, whooping cough, leprosy, insomnia, cancer, gonorrhea, burns, headache, liver disorders, bilious vomiting and sexual asthenia (Deshmukh and Gaikwad, 2013).

Basella alba and *Basella rubra* have been used in traditional medicine system in India for several years (Kumar *et al.*, 2015a, b, c). *B. alba* leaves are used in Ayurveda system of medicine as soothing agents, laxative, demulcent, astringent and the cooked roots are used in the treatment of diarrhoea (Kumar *et al.*, 2013). The plant contains good amount of mucilage and iron. It is diuretic, leaves are demulcent and cooling. It has mild laxative effect. Leaves are reduced to pulp and applied to boils, ulcers and abscesses to hasten suppuration. It is used in utricaria or allergies. Juice of leaves together with sugar-candy (Kalkand) is useful in catarrhal affections of children and administered with much benefit in gonorrhoea and balanits. Leaf juice thoroughly rubbed and mixed with butter is soothing and cooling application for burns and scalds. The decoction of roots of Basella (red type) relieves bilious vomiting.

Therapeutic Properties of Basella

Anti-inflammatory: Arokoyo *et al.*, (2018), investigated the role of *Basella alba* aqueous leaf extract in the modulation of inflammatory cytokines and islet morphology in streptozotocin-induced diabetic rats. From their studies it was concluded that the aqueous extract of *B. alba* stimulates the recovery of beta-islet morphology in streptozotocin induced diabetic rats by modulating the peripheral production of inflammatory cytokines.

Anti-cancerous: The current treatment options of cancer are radiotherapy, chemotherapy, hormone therapy, and surgery, where all of them have unpleasant side effects. Hence, the scientists are trying to seek for noble compounds from natural sources to treat cancer. Islam *et al.*, (2018) conducted an investigation to evaluate *Basella alba* (aqueous extract of leaf and seed) for its antiproliferative effect along with molecular signaling of apoptosis in Ehrlich ascites carcinoma (EAC) cell line. A significant cytotoxic activity was found in both leaf and seed extract. In haemocytometic observation, the leaf and seed extracts exhibited about 62.54±2.41% and 53.96±2.34% cell growth inhibition, respectively, where as standard anticancer drug Bleomycin showed 79.43±1.92% growth inhibition. Morphological alteration under fluorescence microscope showed nuclear condensation and fragmentation which is the sign of apoptosis. Apoptosis induction was also confirmed by DNA laddering in leaf and seed treated EAC cells. Upregulation of the tumor suppressor gene P53 and down regulation of anti apoptotic geneBcl-2 enumerate apoptosis induction. This study also indicated that leaf and seed extracts of *B. alba* have antiproliferative activity against EAC cell line and can be a potent source of anti cancer agents to treat cancer.

Kumar *et al.*, (2018) evaluated the cytoxicity of the extracts of *Basella alba* on A431 (epidermoid carcinoma), Hep G2 (hepatocellular carcinoma) and MG 63 (osteosarcoma) cells and their observations indicated antiproliferative activity against all the cell lines. *B. alba* extract showed higher anti-proliferative activity in the range of 37.95–84.86%. Chick embryo chorioallantoic membrane (CAM) assay revealed inhibition of neo-vessels formation. Significant suppression was found with extracts of *B. alba* at 7 mg/ml compared to that of *B. rubra*. These studies indicate that Basella sps can be used as a source of natural antioxidants and can be of high significance in pharmaceutical and nutraceutical industries.

Cardio & Hepatoprotective: Hypercholesterolemia is the major risk factor that leads to atherosclerosis. Nowadays, alternative treatment using medicinal plants is gaining much attention since the usage of statins leads to adverse health effects, especially liver and muscle toxicity. Baskaran *et al.*, (2015) designed a study to investigate the hypocholestrolemic and antiatherosclerotic effect of *Basella alba (rubra)* using hypercholesterolemia induced rabbits. The treatment with *B. alba extract* significantly lowered the levels of total cholesterol, LDL, and triglycerides and increased HDL and antioxidant enzymes (SOD and GPx) levels. The elevated levels of liver enzymes (AST and ALT) and creatine kinase were noted in hypercholesterolemic and statin treated groups indicating liver and muscle injuries. Treatment with *B. alba* extract also significantly suppressed the aortic plaque formation and reduced the intima: media ratio as observed in simvastatin-treated group. This study on *B. alba* suggests its potential as an alternative therapeutic agent for hypercholesterolemia and atherosclerosis.

References

Arokoyo, D.S., Oyeyipo, I.P., Du Plessis, S.S., Chengou, N.N., and Aboua, Y.G. 2018. Modulation of inflammatory cytokines and islet morphology as therapeutic mechanisms of Basella alba in streptozotocin-induced diabetic rats. Toxicol. Res., (2018), 34(4): 325-332.

Baskaran, G., Salvamani, S., Azlan, A., Ahmad, S.A., Yeap, S.K., and Shukor, M.Y. 2015. Hypocholesterolemic and antiatherosclerotic potential of Basella alba leaf extract in hypercholesterolemia induced rabbits. Evidence-Based Complementary and Alternative Medicine, 2015: Article ID 751714, 7 pages. http://dx.doi.org/10.1155/2015/751714.

Deshmukh, S.A., and Gaikwad, D.K. 2013. A review of the taxonomy, ethnobotany, phytochemistry and pharmacology of Baseslla alba (Basellaceae). Journal of Applied Pharmaceutical Science, (2013),40(01):153-165.

Islam, M.S., Rahi, M.S., Jahangir, C.A., Rahman, M.H., Jerin, I., Amin, R., Hoque, K.M.F., and Reza, M.A. 2018. *In vitro* anticancer activity of Basella alba leaf and seed extracts against Ehrlich's Ascites Carcinoma (EAC) Cell line. Evidence-Based Complementary and Alternative Medicine, 2018: Article ID 1537896, 11 pages. http://doi.org/10.1155/2018/1537896.

Kilari, B.P., Kotakadi, V.S., and Penchalaneni, J. 2016. Anti-proliferative and apoptotic effects of *Basella rubra* (L.) Against 1, 2-Dimethyl Hydrazine-induced colon carcinogenesis in rats. Asian Pac J Cancer Prev., 17:73–80.

Kumar, B.R., Anupam, A., Manchikanti, P., Rameshbabu, A.P., Dasgupta, S., and Dhara, S. 2018. Identification and characterization of bioactive phenolic constituents, anti-proliferative, and anti-angiogenic activity of stem extracts of Basella alba and rubra. J Food Sci Technol., (May,2018), 55(5):1675-1684.

Kumar, S, S., Manoj, P., Shetty, N.P., Prakash, M., and Giridhar, P. 2015b Characterization of major betalain pigments -gomphrenin, betanin and isobetanin from *Basella rubra* L. fruit and evaluation of efficacy as a natural colourant in product (ice cream) development. J Food Sci Technol., 52:4994–5002.

Kumar, S., Prasad, A.K., Iyer, S.V., and Vaidya, S.K. 2013. Systematic pharmacognostical, phytochemical and pharmacological review on an ethno medicinal plant, *Basella alba* L. J Pharmacogn Phytother., 5:53–58.

Kumar, S.S., Manoj, P., and Giridhar, P. 2015c. A method for red-violet pigments extraction from fruits of Malabar spinach (*Basella rubra*) with enhanced antioxidant potential under fermentation. J Food Sci Technol., 52:3037–3043.

Kumar, S.S., Manoj, P., Giridhar, P., Shrivastava, R., and Bharadawaj, M. 2015a. Fruit extracts of *Basella rubra* that are rich in bioactives and betalains exhibit antioxidant activity and cytotoxicity against human cervical carcinoma cells. J Funct Foods., 15:509–515.

10.6. Alternathera (*Alternanthera sessilis*)-Ponnanganni Keerai

This is rightly called "Green Gold" as it has several beneficial effects. The young shoots are nutritious and contain 5% protein and 16.7 mg/100g iron. This green is considered to be galactogogue. If the leaves of this herb are fried in ghee with little pepper and salt and taken for 48 days it promotes the beauty of the body, imparts a golden glow and cools eyes. It also improves production of breast milk in lactating mothers. Eating Ponnanganni leaves cooked with Tuvar dal and ghee increases weight and strength of person. It also improves the function of liver. It also has febri fuge effect (an agent that lessons fever). If eaten with garlic, ponnanganni cures chronic constipation and piles. Taking two ounces of juices of Alternanthra (Ponnangani) plant mixed with 4 ounces of goat or cow's milk reduces the body heat and the body gains strength.

The plant *Alternanthera sessilis* of Amaranthaceae family has been used in ancient India as an abortifacient, galactagogue, cholagogue, febrifuge, as an antidote for snake bite and skin diseases. It is also reported to possess wound healing property (Das, *et al.*, 2015; Paridhavi *et al.*, 2008.), hepatoprotective property, (Das *et al.*, 2015 & 2014) improving memory and intelligence, recuperating skin tone (Roy *et al.*, 2008; Anonymous, 1999; Khare, 2004), eye complaints, (Saqib and Janbaz, 2016; Dangwal *et al.*, 2010), bone fractures and malaria, (Saqib *et al.*, 2016; Kumari *et al.*, 2001), antipyretic (Saqib and Janbaz, 2016; Nayak *et al.*, 2010) and anticarcinogeni (Lalithashree and Vijayalakshmi, 2018.) Among the frequently consumed greens in southern India in general and Tamil Nadu in particular Ponnankanni (Sessile joyweed - *Alternanthera sessilis*) finds an important place. It is also one of the important herbs used in Traditional Siddha Medical system, prevalent in Tamil Nadu. Its Tamil name indicates 'Ponnankanni - Pon aagum kaan nee' (Literally meaning – Your body will get golden luster). According to Traditional Siddha literatures, this herb contains gold and thereby comes under Kaya Kalpa (Panacea) category also. It is said to give five cooling effect to eyes & body, relieves neuritis, treats 96 types of eye diseases and aids disease free healthier life (Walter *et al.*, 2014).

Nutritional Components of *Alternanthra sessilis* L., leaves

Constituents	Quantity per 100g
Moisture	77 g
Carbohydrates	12 g
Proteins	5 g
Fat	0.1 g
Minerals	2 .0 g
Calcium	510 mg
Phosphorus	60 mg
Iron	2 mg
Energy	73 Kcal

Gopalan *et al.*, (1985)

In combination with *Eclipta alba* (Karisilanganni) leaves if it is boiled in gingelly oil and applied to hair and skin it improves the shine of skin and promotes hair growth. Leaves use for increasing hemoglobin, root is used for any swelling.

Phytochemicals

Phytochemical constituents such as flavonoids, tannins, phenols, saponins and several other aromatic compounds are secondary metabolites that were reported in Alternanthera plants that is believed to serve as defence mechanism against microbes, insects and other herbivores (Purkayastha *et al.*, 2005; Magana-Arachchi *et al.*, 2011). The young shoots of *A. sessilis* contain carotenoids, triterpene [Dogra *et al.*, 1977), saponins [Kupundu *et al.*, 1986], flavonoids, steroids, stigmasterol, β-sitosterol [Reni *et al.*, 1975)]. Phytochemical studies yielded β-carotene, ricinoleic acid, myristic, palmitic, stearic, oleic and linoleic acids, α-spiraterol and uronic acid [Reni *et al.*, 1975]. Alternanthera also contains 2, 4- methylene cycloartanol and cycloeucalenol, choline, oleanolic acid. Saponins have been isolated from the leaves. Roots contain lupeol [Leung *et al.*, 1968]. Young shoots contain protein [Gupta, 2004]. It also contains 5-a - stigmasta-7-enol [Ghani, 1996; Rastogi and Mehrothra 1996]. Asolkar *et al.*, 1992 reported the isolation of flavonoids, triterpenoids, steroids and β-sitosterol, stigmasterol, campesterol, lupeol [Sivakumar and Sunmathi, 2016 & Kumar *et al.*, 2011]. Study reported that both aqueous and ethanolic extracts of aerial parts of *A. sessilis* Linn.

Marate and Umate (2018) evaluated phytochemicals present in petroleum ether extract of *A. sessilis* leaves and found presence of about seventeen different phytochemicals using GC-HRMS spectra and identified them with the help of NIIST MS data library. These were reported as oxalic acid (ally nonyl Ester & ally tridecyl ester), hexadecane, 3-Tetradecane, oxirane (tetradecyl), phytol, sulfurous acid, 3-Eicosene (E); Nonadecane, 2-methyl; Eicosane,2-methyl; 1-Docosene; geranylgeraniol; 2-Bromotetradecane; Cholestan-3-ol, 2-methylene; vitamin E(μ tocopherol), tert-Hexadecanethiol and 1-Heptatriacotanol.

Alternanthera sessilis (L.) R.Br. ex DC and *Alternanthera pungens* Kunth are utilized extensively as raw drug sources worldwide in many traditional systems of medicine.

Therapeutic benefits of the *A. sessilis* include anti-inflammatory effect (Subhashini *et al.*, 2010), the nootropic activity (Surendra Kumar *et al.*,2011), cytotoxic effect towards pancreatic cancer cell lines (George *et al.*, 2010), and the free radical-scavenging ability (Borah *et al.*, 2011).The ethanolic extract of *Alternanthera sessilis* Linn. shows a significant antimicrobial activity against microorganisms like *Bacillus polymexia, Salmonella typhii, Candida albicansetc* (Ashik Kumar *et al.*, 2014). Study demonstrated that ASEAF possesses antihyperglycemic effect, anti-triglyceridemic effect, and pancreatic protective effect in obese type 2 diabetic rats (Tan and Kim, 2013). It is a good adjuvant with sex tonics and for females a natural galactagogue (Gangarade *et al.*, 2003; Yadav *et al.*, 2013). Reported the anti diarrhoeal activity of the aqueous extract of dried entireplant material of *A. sessilis*. The hematinic activity of *A. sessilis* (L.) R. Br. was evaluated by monitoring the change in serum ferritin and haemoglobin levels of mice and rats (Erna and Osi, 2010).

The hydroethanolic extract of Alternanthera sessilis was found to have analgesic property with varying potencies when tested in mice using chemical method and hot plate method of pain induction (Mohapatra *et al.*, 2018).

References

Anonymous. 1999. Ayurvedic Pharmacopoeia of India, Part – I, Vol. II, The controller of Publications, Civil lines, Delhi (India). 1999. 104 – 106. 15.

Dangwal, L.R., Sharma, A., Kumar, N., Rana, C.S., and Sharma, U. 2010. Ethno – medico Botany of some aquatic Angiospermae from North – West Himalaya. Researcher, 2: 49 - 54.

Das, M., Ashok Kumar, D., and Das, A. 2014. Hepatoprotective activity of ethanolic extract of *Alternanthera sessilis* linn on paracetamol induced hepatotoxic rats. International Journal of Pharmacy and Pharmaceutical Research, (2014), 4(4): 91 – 96.

Das, M., Ashok Kumar, D., Mastanaiah, K., and Das, A., 2015. Evaluation of anti – diabetic activity of ethanolic extract of *Alternanthera sessilis* linn. In Streptozotocin induced diabetic rats. International Journal of Pharma Sciences and Research, (2015). 6(7): 1027 – 1032.

Gupta, A.K. 2004. Indian Medicinal Plants, ICMR, New Delhi, (2004), 151 – 157.

Khare, C.P. 2004. Encyclopedia of Indian Medicinal Plants, Springer - Verlag Berlin Heidelberg. 2004. 48 -49.

Kumari, B., and Kumar, S. 2001. A check list of some leafy vegetables used by tribals in and around Ranchi, Jharkand. 2001. Zoo's Print J., 16:442 – 444.

Lalithashree, T., and Vijayalakhsmi, K. 2018. Evaluation of vitamins and antinutrients in the leaves of traditional medicinal plant *Alternanthera Sessilis* (L.) R.Br.Ex DC, International Jounral of Health Science and Research, (2018), 8(10): 244-253.

Nayak, P., Nayak, S, Kar, D.M., and Das, P. 2010. Pharmacological evaluation of ethanolic extracts of the plant *Alternanthera sessilis* against temperature regulation. J. Pharm. Res., 3(6): 1381 – 1383.

Paridhavi, P., Sunil, S.J., Nitin, A., Patil, M.B., Chimkode, R., and Tripathi, A. 2008. Antimicrobial and wound healing activities of leaves of *Alternanthera sessilis*. Int J of Green Pharm., (2008), 2: 141-144.

Roy, A., and Saraf, S. 2008. Antioxidant and Antiulcer activities of an ethnomedicine: *Alternanthera sessilis*. Research J. Pharm and Tech., (2008). 1(12):1-6.

Saqib, F., and and Janbaz. 2016. K.H., Rationalizing ethnopharmacological uses of *Alternanthera sessilis*: A folk medicinal plant of Pakistan to manage diarrhoea, asthma and hypertension. Journal of Ethnopharmacology, (2016), 182: 110 -121.

Walter, T.M., Perish, S., and Tamizhamuthu, M. 2014. Review of *Alternanthera sessilis* with reference to traditional Siddha medicine. International Journal of Pharmacognosy and Phytochemical Research, (2014), 6(2): 249-254.

Purkayastha, J., Nath, S.C., and Islam, M. 2005. Ethnobotany of medicinal plants from Dibru-Saikhowa Biosphere Reserve of Northeast India. Fitoterapia, 76: 121-127.

Magana-Arachchi, D.N., Medagedara, D., and Thevanesam, V. 2011. Molecular characterization of *Mycobacterium tuberculosis* isolates from Kandy, Sri Lanka. Asian Pac J Trop Dis., 1: 181-186.

Dogra, J.V.V., Jha, O.P., and Mishra, A. 1977. Cheamotaxonomy of Amaranthaceae - study of triterpenes. Biochem J., 4: 14-18.

Kapundu, R., Mpuza, L., Nzunzu, D. 1986. A Saponin from *Alternanthera sessilis*. Bul. Soc. R. Sci. Liege, (1986), 55(5-6): 605-66.

Shinne Reni, Chen, Arh- Hwang. 1975. Kaohsung Medical college Kaohsung (Taiwan) Tai_ Wan_yao Hsueh Tsa chils. 1975; 27(1-2): 103-4.

Leung, W.T.W., Busson, F, and Jardin, C. 1968. Food composition table for use in Africa. FAO, Rome, Italy, 1968; 306.

Gupta, A.K. 2004. Indian medicinal plants New Delhi. ICMAR 151-7.

Ghani. 1996. Medicinal plants of Bangladesh: Chemical constituents and uses, 2nd Ed Asiatic Society of Bangladesh 1-17.

Rastogi, R.P., and Mehrotra, B.N. 1996. Compendium of Indian Medicinal Plants. Publications & Information Directorate CSIR New Delhi India 5: 224.

Asolkar, L.V., Kakkar, K.K., Chakre, O.J. 1992. Secondary Supplement to Glossary of Indian Medicinal Plants with Active Principles. Publications and Information Directorate (CSIR), Dr. K.S. Krishnan Marg New Delhi 332-335.

Sivakumar, R., and Sunmathi, D. 2016. Phytochemical screening and antimicrobial activity of ethanolic leaf extract of *Alternanthera sessilis* (l.) r.br. ex dc and Alternanthera philoxeroides (mart.) griseb, ejpmr 3: 409-412.

Surendra Kumar, M., Silpa Rani, G. 2011. Screening of Aqueous and Ethanolic Extracts of Aerial Parts of *Alternanthera sessilis* Linn. R.br.ex. dc for Nootropic Activity. J Pharm Sci Res, 3: 1294-1297.

Subhashini, T., Krishnaveni, B., and Reddy, S. C. 2010. Anti-inflammatory activity of the leaf extract of *Alternanthera sessilis*, HYGEIA Journal for Drugs and Medicines 2: 54–57.

Surendra Kumar, M., Silpa Rani, G., Swaroop Kumar. 2011. Screening of aqueous and ethanolic extracts of aerial parts of *Alternanthera sessilis* Linn. R.br.ex. dc for nootropic activity. Int J Pharm Sci Res., 3: 6 1294-1297.

George, S., Bhalerao, S.V., and Lidstone, E.A. 2010. Cytotoxicity screening of Bangladeshi medicinal plant extracts on pancreatic cancer cells. BMC Complementary and Alternative Medicine, 10: 52.

Borah, A., Yadav, R.N.S., and Unni, B.G. 2011. Invitro antioxidant and free radical scavenging activity of *Alternanthera sessilis*. Int J Pharm Sci Res., 2: 1502–1506.

Ashok Kumar, D., Das, M., Mohanraj, P. 2014. Antimicrobial activity study of ethanolic extract of *Alternanthera sessilis*, Linn Aerial parts. Journal of Applied Pharmaceutical Research, 11: 1-4.

Gangrade, S.K., Tripathi, N.K., and Harinkhede, D.K. 2003. Ethnomedicinal diversity used by tribals of Central India. Sam Com Digital Graphics Indore, India 19-20.

Erna C, Arollado., and Marina, Osi. 2010. Hematinic activity of *Alternanthera Sessilis* (L.) R. Br. (Amaranthaceae) In Mice and Rats. E-International Scientific Research Journal, II: 2094-1749.

Nayak, P., Nayak, S., and Kar, D.M. 2014.Pharmacological evaluation of ethanolic extracts of the plant *Alternanthera sessilis* against temperature regulation. JPR, 2014; 3(6): 1381-1383.

Mohapatra, S.S., Kafle, A., Chatterjee, J., Mohan, P., Roy, R.K., Reddy, I. 2018. Analgesic activity of hydroethanolic extract of *Alternanthera sessilis* in mice. Journal of Pharmacognosy and Phytochemistry 2018; 7(4): 1836-1839.

Tan, Kok Keong., and Kim, Kah Hwi. 2013. *A. sessilis* red ethyl acetate fraction exhibits antidiabetic potential on obese type 2 diabetic rats. Evidence-Based Complementary and Alternative Medicine, 2013: 1-8.

Yadav, Sanjay Kumar., and Das, Sanjib. 2013. Evaluation of anti-diarrhoeal property of crude aqueous extract of *Alternanthera sessilis* Linn. International J Pharm Innov., (2013), 3(3):110-115.

10.7. Black Nightshade (*Solanum nigrum* L.)

Commonly known as Black nightshade, *Solanum nigrum* Linn. (Manathakkali Keerai in Tamil) is a garden weed. The plant is cultivated as a leaf crop in some areas. This leafy green is very popular in Tamil Nadu, and its fruits look like small and tiny tomato. These fruits are green or purple in color and its contents are bitter. There are no specific varieties listed anywhere. It has been used in early Ayurveda along with other ingredients in heart disease. The active principal constituent is **solanarigine** and **solanine**. The plant is effective in treatment of cirrhosis of the liver. The plant is also credited with emollient, diuretic, antiseptic and laxative properties. The fruits and leaf dish are common in Tamil Nadu and other southern states. In north India, the boiled extract of leaves and berries are used to alleviate discomfort in liver related ailments, including jaundice.

Indian Meteria Medica described this herb as sedative, diaphoretic, diuretic, hydrogogue and expectorant. Solanine is a powerful protoplasmic poison acting upon amoeboid organisms and ciliated epithelial cells. The berries are tonic, diuretic and useful in heart diseases.

Nutritional Components of *Solanum nigrum* L.

Constituents	Quantity (per 100g)
Moisture	82 g
Carbohydrates	9 g
Proteins	6 g
Fat	0.1 g
Minerals	2 g
Calcium	410 mg
Phosphorus	70 mg
Iron	20 mg
Energy	68 Kcal

Gopalan *et al.*, (1985)

As a medicinal herb, leaves of *Solaneum nigrum* are employed as poultice over rheumatic and gouty joints and also a remedy in skin diseases. Freshly prepared fluid extract from all portions of plant are recommended in dropsy, piles, gonorrhoea, inflammatory swellings and chronic cirrhosis of lever and spleen. A syrup made from fruit and plant is useful as a cooling drink in fevers and to promote perspiration. The leaves are made hot and applied to painful and swollen testicles.

10.8. Agathi Keeai (*Sesbania grandiflora*)

The tender leaves, green fruit and flowers are eaten alone as a vegetable or mixed into curries or salads. Agathi is a folk remedy for diuretic, emetic, emmeuagogere, febrifuge (fever) conditions. It is also useful for healing bruises, catarrh, dysentery and head aches. It is also reported to be useful for treating sores, sore throat and stomatitis. The bark of the tree is astringent and is used in treating fever associated with smallpox. The juice from the flowers is used to treat headache, head congestion or snuffy nose. As a snuff the juice is supposed to clear nasal sinuses. Ayurvedics believe that the leaves are useful as alexeteric, anthelminthic, epilepsy, gout, itch, leprosy, nyctalopa and opthalmia. Yunani considers the leaf tonic useful in biliousness, fever and nyctalopia. Crushed leaves are applied to sprain and contusions. A gargle with leaf juice cleanses the mouth and throat.

Nutritional Components of Agathi (Sesbania grandiflora) leaves

Constituents	Quantity per 100g
Moisture	73 g
Carbohydrates	12 g
Proteins	8 g
Fat	0.1 g
Minerals	3.0 g
Calcium	1130 mg
Phosphorus	80 mg
Iron	4 mg
Energy	93 Kcal

References

Bose, T.K. and Som, M.G. 1986. Vegetable crops in India. Naya Prokash, 206, Bidhan Sarani, Calcutta.

Gopalakrishnan. 2007. Vegetable Crops. New India Publishing Agency. New Delhi. India. Pp.304.

Nadkarni, A.K. 1976. Indian Materia Mecica. Indian Council of Scientific and Industrial Research. New Delhi.

Thomas Paul A. Devasagayam.2007. Indian Herbs and Herbal Drugs Used for the Treatment of Diabetes. J. Clin. Biochem. Nutr., 40, 163–173, May 2007.

Lightning Source UK Ltd.
Milton Keynes UK
UKHW020103230722
406270UK00004B/278